D0953271

LIFE IN DARWIN'S UNIVERSE

Books by Gene Bylinsky

LIFE IN DARWIN'S UNIVERSE

MOOD CONTROL

THE INNOVATION MILLIONAIRES

CONCORDIA UNIVERSITY LIBRARY
2811 NE HOLMAN ST.
PORTLAND, OR 97211-6099

GENE BYLINSKY

LIFE IN DARWIN'S UNIVERSE

Evolution and the Cosmos

Illustrated by

Wayne McLoughlin

DOUBLEDAY & COMPANY, INC.
GARDEN CITY, NEW YORK
1981

Library of Congress Cataloging in Publication Data

Bylinsky, Gene.
Life in Darwin's universe.

Includes index.
1. Evolution. 2. Life on other planets. 3. Life—Origin.
I. McLoughlin, Wayne. II. Title.
QH371.B86   577
ISBN: 0-385-17049-1
Library of Congress Catalog Card Number 80–2988

Copyright © 1981 by GENE BYLINSKY
Illustrations copyright © 1981 by WAYNE McLOUGHLIN
All rights reserved
Printed in the United States of America

BOOK DESIGN BY BENTE HAMANN

First Edition

Dedicated with thanks to
N. John Berrill, George Gaylord Simpson, Cyril Ponnamperuma,
Frank D. Drake, David Black, and all the other seekers
who have opened the universe of life for us

ACKNOWLEDGMENTS

Much like the evolution of life, the evolution of this book began long ago. In a way, the idea was always in my mind. Bob Guccione and Kathy Keeton gave it a big push forward by enthusiastically responding to my suggestion that I do an article on life-forms elsewhere for *Omni* magazine. That brought me in contact with Wayne McLoughlin, a talented young artist. A kindred soul, Barry Lippman, an editor at Doubleday, spotted the possibilities for a book, and my agent, Diane Cleaver, was in the right place at the right time to agree. The result owes much to the many scientists who gave generously of their time, to Wayne McLoughlin's imagination and skill in illustrating the book, to Barry Lippman's incisive and sensitive editing, and to all the help from Gwen and Carmela. Jackie McLoughlin deserves credit for helping originate the design of the cover. Last, but not least, thanks are due to Bill Rukeyser and Dick Armstrong, of FORTUNE magazine, for graciously consenting to the leave of absence that allowed me to write this book.

CONTENTS

INTRODUCTION

The possibility of life elsewhere in the universe has intrigued and excited man since time immemorial. As long ago as A.D. 1200, the Chinese philosopher Teng Mu expressed this idea poetically when he wrote: "How unreasonable it would be to suppose that besides the Earth and the sky which we see, there are no other skies and no other Earths."

As for the types of life elsewhere, more often than not they were visualized as being unlike any on Earth. In keeping with the oddity theme, science fiction writers and Hollywood movie makers built up a whole menagerie of space monsters. And at that, extraterrestrial life was said to be exceedingly rare.

Lately, however, something new and much more exciting has been happening. A tandem realization began to take hold, first, that life is widespread, and second, that it is much more like ours than anyone ever imagined—although not an exact copy, to be sure.

Support for the conclusion that life is common throughout the universe comes from recent discoveries by radio astronomers that the basic molecules that make up life are scattered throughout the velvety blackness of the cosmos like seeds on a freshly plowed field. As a result of these startling observations, we now know that the universe is a vast chemical laboratory where not only is the matter that goes into the making of the stars the same everywhere, but the building blocks of life are also the same. It's a discovery whose full impact is just beginning to be felt.

The realization that Darwinian evolution prevails throughout the cosmos comes from new discoveries closer to home: from new findings about the evolution of life on Earth. It is this developing idea that is elaborated on in this book to show, in pictures and in words, what life is likely to be like amid the stars.

To lead to that picture of life in the cosmos, we must sketch in the mosaic of the foundations of life on Earth. As C. Owen Lovejoy, a professor of anthropology at Kent State University, has succinctly put this idea in a paper presented at a recent NASA symposium: "If we wish to make estimates of the probability of intelligent life on suitable planets, then we must clearly identify the events and processes by which it appeared on this planet. This is the only method available short of fantasy."

Accordingly, this book begins at the beginning, the creation of our universe in the immense explosion that has become known as the Big Bang. Subsequent chapters trace the chemical and biological evolution of life on Earth and transport the reader to other settings to view the possibilities bound to exist on the billions of life-bearing planets that are believed to circle other suns. The concluding chapters consider the chances of locating other life-bearing planets and establishing contact with their inhabitants.

It is the growing realization that life everywhere must obey the universal laws of chemistry and physics that allows us to proceed with confidence in drawing a new picture of a Darwinian universe. More and more, scientists who study life on Earth are reaching that conclusion about the nature of life throughout the universe.

As Dr. Lovejoy has pointed out, it is reasonable to assume that organic evolution is a process that obeys the same fundamental rules throughout the cosmos as it has on Earth; even if we cannot observe the actual life-forms elsewhere, we can make reasonable inferences about ways life may evolve.

This book, then, is an illustrated travel guide to life in Darwin's universe. It begins, as all journeys must, with the first step of our immense journey: the foundations of the creation of life.

G.B.

LIFE IN DARWIN'S UNIVERSE

Part I

IN THE BEGINNING...

CHAPTER ONE

Building Life's Stage

The universe was born in a blinding flash that lit up the blackness and continued to shine for nearly a million years. "In the beginning was the word . . . ," says the Bible, and the word, say the cosmologists, was light.

Actually, there was a beginning before the beginning. Somehow, prior to that big blast that occurred 15 to 20 billion years ago, the blast that has become known as the Big Bang, matter—perhaps the remains of a previous universe that had contracted onto itself—had condensed into one primordial superatom. The explosion was unlike any we are familiar with on Earth. Instead of starting from a center, it occurred everywhere at once, making every particle rush away from every other particle.

So hot was this fireball that matter arising in the explosion consisted not of familiar molecules and atoms but of various types of elementary particles, those fragments of matter that are being studied today in huge particle accelerators by high-energy physicists. Among particles present in large numbers were the electrons that make up outer shells of atoms. Electrons are negatively charged. Another common particle was the positively charged counterpart of the electron, the positron. The fireball also produced many kinds of neutrinos, the ghostly particles, with little or no mass or electric charge, that can—and do—slice through the Earth (indeed, the universe) as if it weren't there.

There were also a few heavier particles—neutrons and protons—
around. When the temperature dropped, after the first few minutes,
protons and neutrons began to form into complex nuclei. First, the
nucleus of heavy hydrogen (deuterium) was created; it consists of
one proton and one neutron. Next, some of the nuclei assembled to
make the nucleus of helium, which consists of two protons and two
neutrons. The ratio was about 73 percent heavy hydrogen and 27
percent helium nuclei. There were not yet *atoms* of hydrogen and
helium. It took some 700,000 years—a short time on the cosmic scale
—for matter to become cool enough for electrons to join with nuclei
to actually form stable atoms of hydrogen and helium.

Heavier atoms still couldn't arise to any appreciable extent because
the radiation that filled the universe blasted such heavier nuclei apart
as fast as they formed. At that point, the universe consisted mainly of
hydrogen and helium nuclei, and continuous radiation made the universe as bright as a sunny day on Earth. Gradually, gravitation began
to twist hydrogen and helium gases into clumps that ultimately condensed into galaxies and stars. Stars formed through gravitational instability. It's a simple physical concept that was first conceived by
Isaac Newton in the seventeenth century. Such instabilities arise
when a uniform, stable cloud of gas in space is somehow disturbed—
perhaps by shock waves from explosions of other stars. Thanks to
this disturbance, one small spherical region in the gas cloud becomes
somewhat denser than the gas surrounding it. The denser region's
gravitational field becomes somewhat stronger too. It gradually at-
tracts more and more matter and contracts until the internal temperature of the condensing matter ignites the star fires—the thermonuclear reactions. A star is born. Deep inside solar furnaces, heavier
elements begin to be cooked, notably carbon and iron. The first stars
lit up about 200 million years after the Big Bang. Our sun was not
among them. In fact, on the cosmic scale it is a relative youngster,
being approximately 5 billion years old.

In our galaxy, stars range in mass from one hundredth the mass of
the sun to about one hundred solar masses. The little stars are the
longest-lived ones. They shine for tens of billions of years, then fade
slowly to become "white dwarfs" or even denser little stars whose
light eventually goes out and leaves a cold cinder. The massive stars,
on the other hand, live for only a few million to tens of millions of
years. They end their lives in spectacular explosions: the supernovae.

The tremendous force of the supernova explosion forges elements heavier than iron and blasts as much as 90 percent of the big star's material into interstellar space. Out of this "star dust," it is now clear, are born new stars, like our sun, and planets like the Earth. In fact, we and the rest of life on Earth owe our existence to the fact that massive stars have lived and died. Were it not for the supernova explosions, carbon—the key to life—would not be distributed as widely through interstellar space as it is now known to be.

How solar systems such as ours were formed is known in general outline. A spherical blob of hydrogen and helium got flattened into a disk; in the middle of the disk formed the sun, and in the middle and on the edges of the disk formed the planets, as shown in the illustration on the next page.

That's the sketchiest of outlines, of course. Scientists still have some trouble explaining why the so-called terrestrial, or Earth-like, planets—Mars, Venus, Mercury—differ so markedly from the outer giants known as the Jovian planets—Jupiter, Saturn, Uranus, and Neptune—which consist primarily of hydrogen, carbon, nitrogen, and oxygen. Not long ago, a theory was in vogue that the Earth and other planets were literally torn from the sun by a star that passed close to the sun. The material torn from the sun supposedly was tossed into orbit around the sun and, after cooling, condensed into planets. This theory, however, has been abandoned for a number of reasons. For one thing, scientists now think that it's extremely unlikely that a star could have passed so close to the sun—within a few diameters of the sun. With the nearest neighbor star 25 trillion miles away, and the sun's diameter a mere million miles, the chances of such collisions are extremely small. Furthermore, a closer look shows that any suitable pieces of material ripped from the sun would have been forced into hyperbolic, rather than elliptical, orbits, in effect hurled into space, away from any projected solar system.

Today's thinking tends to favor instead a supernova explosion as a trigger of planetary-system formation. This dying blast of a huge star is so powerful that shock waves from it compress the interstellar gas and make it condense into solar systems. The supernova theory, first proposed by Fred Hoyle and Ernst Öpik in the 1940s, has since then been buttressed by a number of experiments and observations. Formation of new stars has in fact been observed on the fringes of shock waves produced by supernovae. Young sunlike stars examined in such locales show the presence of extrastellar material—matter that came from the outside. And in meteorites, scientists recently have

been finding evidence that a supernova-like explosion did indeed take place in our vicinity at the time of the solar system's formation, around 4.5 billion years ago.

Much of the evidence has been gathered by Dr. Gerald J. Wasserburg and his associates at the California Institute of Technology. The evidence places the birth of the solar system and the explosion no more than 5 million years apart, an insignificant time difference in cosmic terms.

It appears that when the big star—probably twenty times the size of the sun—blew up, it made a whole spectrum of radioactive elements. Some of these elements—actually their isotopes, or slightly different chemical forms of the same element—were found by Dr. Wasserburg and other scientists in two kinds of meteorites, both of which scattered debris across Mexico. Inside the first, the so-called Allende meteorite, scientists found tiny spheres containing radioactive elements in amounts much larger than any ever found on Earth or even in moon rocks, which are about a billion years older than the oldest rocks found on Earth. (Lack of erosion preserved older rocks on the surface of the moon.)

The Allende meteorite contained strange mixtures of barium, calcium, strontium, neodymium, and samarium. Found in particularly large amounts was an isotope called magnesium 26, which is produced only by the radioactive decay of another isotope, aluminum 26. This evidence was intriguing, for aluminum 26, because of its relatively short half-life of seven hundred thousand years, has all but vanished from the solar system. (Half-life is the rate of decay of radioactive materials, the time in which half of a given number of atoms in the material disintegrate.)

The evidence furthermore meant that a hydrogen-bomb-like explosion, but on an immensely vaster scale, had produced aluminum 26, and that enormous heat was being generated inside the rocks from the decay of radioactive isotopes of aluminum and other elements. More supporting evidence came from the second stone that fell from the sky, the so-called Santa Clara meteorite. There, an unexpected amount of an isotope known as silver 107 was discovered. It was produced by the radioactive decay of palladium 107, whose origins date back 4.6 billion years. Palladium can be made only inside a nuclear reactor—or in the nuclear furnace of an exploding star.

Early in 1980, one of Wasserburg's associates, Franz Niederer, found an unusual form of titanium in the "star dust" particles scooped up high above the atmosphere by a NASA U-2 plane. He dated this type of titanium at 2 to 3 million years older than the oldest material so far recovered in the solar system, the 4.5-billion-year-old moon rocks. This finding also supports the supernova theory: the special form of titanium was apparently formed in the solar system's parent supernova.

Would such supernova explosions make solar systems of our type rare? Not particularly. One or two supernova blasts are observed every two centuries or so in galaxies similar to ours. Considering the size and the age of the universe, these events would still add up to an astronomical number. Conditions that lead to the formation of a solar-type system thus may be special but not necessarily rare.

There is still no universal agreement, though, on how planets were formed in our solar system. The body of the ongoing work in this area suggests that some of the material collapses because of gravitational instability into protoplanets, giant forerunners of planets. At least that's how the gas giants such as Saturn and Jupiter are believed to have formed. Terrestrial planets, on the other hand, probably formed by aggregation of a succession of different-sized solid bodies, starting with grains of dust that collected into *planetesimals*—bodies of up to six hundred miles wide—which, in turn, aggregated into today's terrestrial planets. It's possible that one process dominated at an early stage of planet formation, while another played a major role at a later time.

Still another possibility—a sudden flareup by the sun that blew hydrogen and helium away from the vicinity of the sun—would account for the differences in structure between the terrestrial and the Jovian planets. The solar wind would have blown away just the light gas particles but not the heavier grains of dust.

The planetesimal theory accounts for the depletion of the chemically nonreactive noble gases argon and neon, since they would not have been retained by small objects with low gravity. Furthermore, the gases would be lost when the remaining nebula, or cloud of gas, dissipated.

The larger-scale gravitational collapse, on the other hand, explains why the giant planets have satellites with solid surfaces: the giant planets and their satellites constitute miniature solar systems, in which the two processes were at work just as they were in the solar system as a whole.

To be sure, many scientists find the idea of two fundamentally different processes having taken place in the solar nebula unpalatable, so that there is not yet a unified, well-accepted theory of planetary formation.

The planets and their atmospheres—at least their original atmospheres—should reflect the cosmic ratios of the elements. Everywhere we look, we see stars, galaxies, clouds of gas in space—all composed of the same basic elements in the same proportions. By volume, hydrogen accounts for 90 percent of the total visible universe, helium for more than 9 percent, and all the other elements for less than 1 percent, with oxygen, carbon, neon, and nitrogen being the most common.

Accordingly, the planets' primeval atmospheres should have contained mostly hydrogen, along with helium, oxygen, carbon, neon, and nitrogen. Helium and neon are inert, but carbon, nitrogen, and oxygen readily combine with hydrogen, forming methane (CH_4), ammonia (NH_3), and water (H_2O). That type of atmosphere closely reflects the composition of the sun, and in fact, the atmospheres of Jupiter and the other giant planets are dominated by hydrogen, with trace amounts of methane and ammonia. Helium and water are also assumed to be there.

In contrast, the Earth and the other terrestrial planets, with their weaker gravitational pull, are believed to have lost their primeval, hydrogen-dominated atmospheres. Hydrogen and helium, the two lightest elements, escaped into space, while heavier elements remained behind. Methane was oxidized to CO_2 and ammonia to molecular nitrogen. Today the Earth's atmosphere is made up of 78 percent nitrogen, 21 percent oxygen, and 1 percent other gases, with only trace amounts of hydrogen and carbon dioxide.

There is geological evidence that the composition of the Earth's atmosphere was quite different in the remote past. Certain rare minerals, apparently produced in a hydrogen-rich atmosphere, change in chemical composition when exposed to oxygen in the atmosphere. For example, a mineral called uraninite, formed 2 to 3 billion years ago, turns into a type of uranium oxide when it is brought into the open air.

The creative processes by which "the present beautiful order of the solar system was developed," as T. J. J. See, a turn-of-the-century American astronomer put it, of necessity involved violence. For one thing, tremendous impacts by smaller bodies cratered the Earth—as they so tellingly cratered the moon.

But those ugly pockmarks were obliterated on Earth by developing atmosphere and oceans. Both the atmosphere and the oceans owe their origins to violent upheavals inside the forming Earth. Volcanoes spewed out gases that went into the making of an atmosphere. A mixture of carbon dioxide, methane, carbon monoxide, ammonia, water, nitrogen, and hydrogen escaped to the surface and became part of the primitive atmosphere. Those gases also carried water with them; it collected in the depressions in the crust to form ponds at first, then lakes, and then oceans, as shown in our illustration of the primitive Earth.

Ultraviolet radiation from the sun beat down on the oceans' surface unhindered by any protective layer of ozone, since the Earthly atmosphere consisted mainly of those original constituents of the universe: hydrogen and helium. With water plentiful and radiation acting as a catalyst, the stage was set for the emergence of life.

CHAPTER TWO

The Curtain Rises

The primitive Earth on which life arose would hardly make an attractive place for us to visit, much less to live on. Sinister lightning flashed and volcanoes spewed poisonous gases into the atmosphere—a hydrogen-rich, or "reducing," atmosphere in which man and other oxygen-breathing creatures could not survive.

Yet the Earth obviously was the kind of planet needed for life to emerge—or at least for life as we know it, as scientists like to qualify that statement. Primitive forms of life, most scientists now believe, could only have arisen in a reducing atmosphere. Reactions that lead to the formation of prebiotic molecules—formation of nucleotides and amino acids, which are the building blocks of living matter—can proceed only in a reducing atmosphere. Molecular oxygen is a poison to organic, or carbon-based, molecules; it decomposes them. Butter turning rancid when left out in the open warm air is an example of such decomposition.

Paradoxically, life began its explosive expansion that eventually led to man in an oxygen-rich, or "oxidizing," atmosphere. Organisms had to adjust to living with "poisonous" oxygen and learn to like it. If life is to become more complex than consisting merely of organisms containing a few cells, if multicellular organisms are to emerge and thrive, they must become oxygen breathers. The reason lies in the realities of energy production.

So much more efficient is energy production by oxidation of glucose to CO_2 (carbon dioxide) and water, for instance, that thirty-six adenosine triphosphate (ATP) molecules are produced, as contrasted with merely two molecules of ATP that can be obtained by fermentation in the absence of oxygen. ATP is the universal energy currency of the cell. Its uses range from making our muscles move to igniting the firefly's lantern. (ATP produces light by transferring part of its energy to a protein that changes shape and emits light as it does so.)

This fact about biological energy production has a direct bearing on the types of creatures that are likely to exist on other planets. If those life-forms have gone beyond the stage of bacteria, they have to breathe oxygen, hinting at a resemblance to us, at least in that respect.

When life began on Earth, though, there wasn't any oxygen in the atmosphere. The first organisms, presumably fermenting bacteria, exploited organic, or carbon-containing, molecules available in the environment. They ate up the organic "soup." This is another reason why, aside from the absence of a reducing atmosphere, life can't arise in a present-day setting; such primitive organisms would be quickly consumed by the more advanced ones.

It's generally agreed that the primitive atmosphere was reducing, although scientists disagree about its exact composition. Those geologically inclined tend to favor a carbon monoxide-carbon dioxide atmosphere. Chemists, on the other hand, believe that the atmosphere contained more hydrogen, because carbon-based building blocks of life can be made more readily in such an atmosphere. The geologic and paleontologic records offer no decisive data on this question.

What is clear, however, is that the basic building blocks of life are easy to produce from nonliving matter. Stanley Miller showed this when he was a graduate student at the University of Chicago in 1953. Reconstructing the Earth's primitive atmosphere, consisting of methane, ammonia, and water vapor, in a laboratory flask, and using spark discharges and other devices as energy sources, Miller and other scientists elsewhere eventually produced most of the components of the basic molecules that constitute life: nucleotides and amino acids.

Those two basic molecules go, respectively, into the making of the information-storing DNA (deoxyribonucleic acid), and the information-transcribing RNA (ribonucleic acid), and make up proteins which perform both functional and structural roles. Nucleic acids and proteins make up all living creatures, from an elephant to an *Escherichia coli* bacterium. The nucleic acid molecule consists of nucleotides strung together like beads in a necklace. In turn, the nucleotides are made up of a purine or a pyrimidine "base," a sugar, and a phosphate. The protein molecule, on the other hand, is made up of twenty amino acids, which link up with one another in various combinations to form large molecules.

Stanley Miller is now a professor at the University of California at San Diego, where he continues his research. Says Dr. Miller: "Now there is a whole body of work showing that organic compounds are easy to make and the ones you get are those—more or less—that occur in living organisms. You *don't* make a little bit of everything. You tend to concentrate on the ones that are prominent in living organisms."

In those early experiments, four amino acids were produced. What's more, later work showed that the amino acids in the primordial soup had already formed chains of molecules (called polymers), suggesting that chemical evolution of the first proteins may have required a relatively short time. One of the random chains could have acted as a catalyst and promoted the generation of nucleic acids. The structural components of the nucleic acids (the purines adenine and guanine) are easily synthesized from hydrogen cyanide, which is the primary product of experiments with primitive atmospheres and various radiation and heat sources. Sugars, for their part, can be made from formaldehyde, which is also easily generated in experiments. So those forbidding substances—hydrogen cyanide in today's atmosphere would be lethal to life, and formaldehyde is also associated with dead rather than living matter—in combination give rise to the functional dynamos of life.

Heating those building blocks with inorganic phosphates, furthermore, forms nucleotides, the more complex units of nucleic acids. It's a simple reaction that could have easily occurred on the prebiotic Earth.

Since the original Miller experiment, Dr. Cyril Ponnamperuma, a prominent student of the origins of life and professor at the University of Maryland, has carried this research to the point where he can see patterns emerging by which differing elementary molecules recognize each other. This is a basic step toward producing self-replicating molecules in the laboratory, and that's how the original molecules came together to form the first organisms.

"We can make molecules necessary for life—that's been done," says Dr. Ponnamperuma. "If I can demonstrate a replicating molecule, I'll die a happy man."

Such an achievement would show, of course, as the Ceylon-born scientist puts it, "If I can do it in my lab flask, there may be thousands of such natural flasks around the universe."

On Earth, the coming together of those ingredients to produce the first self-replicating molecule (an example of which is seen on the opposite page) was the crucial moment, a miracle of chemical creation.

That's one way life could have arisen. Another way would have been for a prototype cell to arise by forming a rudimentary membrane to protect itself against environmental trauma. In that case, the genetic machinery and the structural components would have developed simultaneously.

Which came first, the cell or the gene?

Scientists still argue the question. The great Russian biochemist Aleksandr I. Oparin, who first proposed the chemical origins of life in 1925, has always leaned toward the cell—a metabolic unit protected by a simple membrane against dilution and damage by its surroundings. Oparin felt that the cell rearranged the atoms it absorbed into molecules needed for the maintenance of the cellular structure and function. And it survived by reproducing itself. To Oparin, reproductive machinery and DNA are the fine parts of the cellular mechanism that turned nonliving forerunners of cells, or protobionts, into living cells.

The great British biochemist J. B. S. Haldane, who published ideas similar to Oparin's in 1929, on the other hand, favored the "naked gene"—a large DNA molecule that made copies of itself by utilizing materials in its surroundings.

Oparin's idea so far has been easier to support, up to a point, through laboratory experiments. In those tests, it has been shown that simple chemical reactions can proceed in droplets formed of proteins and placed in an appropriate medium.

Oparin has been looking for many years into an interesting phenomenon that may suggest how the first cells formed. This is the tendency of polymers, or long-chained molecules, to separate into colloidal droplets called coacervates when suspended in a water-rich medium.

Not only that, but many of these droplets display a type of a membrane, a vital ingredient of any cell, of course.

Interestingly enough, Oparin and his associates found that the way to stabilize the droplets for hours, and even days and weeks, is to give them a rudimentary kind of metabolism. Oparin had discovered, for instance, that when he added a particular enzyme (phosphorylase) to a solution containing a protein (histone) and a carbohydrate (gum arabic), the enzyme would become concentrated within the densely clustered droplets.

Furthermore, when glucose phosphate was dissolved in the water, it entered the droplets and was then turned into starch by the enzyme. This increased the bulk of each droplet and caused it to break up into a number of daughter droplets. Those daughter droplets that got a sufficient amount of enzyme continued to grow, although at a slower rate than their "mother" droplet. If these droplets could make more enzyme—which they can't—they would become an example of prototypes of a cell with a simple kind of self-perpetuating metabolism. This isn't creation of living systems, of course, but a hint at how they might have arisen.

In contrast, no one has been able to show yet in the laboratory how DNA can arise from random chemical subunits.

Actually, scientists suspect that all cellular components evolved in parallel. So closely interwoven are cellular activities that scientists have trouble visualizing a simpler system. Nucleic acids can't reproduce themselves without enzymes, those busy catalysts of cellular reactions, and enzymes can't be made without nucleic acids.

Possibly, DNA and proteins evolved together in close collaboration. The nucleic acid could have served as a template, or assembler, of long-chained molecules, and the protein molecules, in turn, protected the DNA by serving as membranes for the prototype cells. The biochemical abilities of the parent cell were passed on to the daughter cells. This machinery was subject to internal error and to externally induced changes in the genetic code—mutations by radiation and other environmental agents. Those organisms with altered genes that could cope best with the environment survived, beginning the long march of natural selection, or evolution as Charles Darwin defined it.

It appears, though, that even then, billions of years ago, life was governed by energy crises. The first microorganisms evolved together with the atmosphere. They influenced its composition by their metabolic development, and the changes in the atmosphere, in turn, stimulated the evolution of new metabolic capabilities. Those first organisms most likely were fermenting ones, living off carbon-containing molecules. These microorganisms multiplied, but their growth became limited by the limited energy derived from fermentation. That was the first energy crisis.

To exploit a new energy source, the chemical energy contained in the mixture of carbon dioxide and hydrogen in the primitive atmosphere, new types of bacteria evolved. Eventually, though, these microorganisms decreased the level of atmospheric hydrogen to a point where further extraction of chemical energy from the atmosphere became difficult. A second energy crisis loomed.

It was solved by the evolution of photosynthesis, the intricate process that allows bacteria (and green plants) to absorb a photon, or a light particle from the sun, and to employ the photon's energy to drive the reaction that transforms sunlight, carbon dioxide, and water into chemical energy stored as starches and simple sugars.

The process also releases oxygen into the air by splitting the water molecule. Photosynthesis, therefore, can justifiably be called the most important biochemical event to evolve on Earth, for it led eventually to the development of oxygen-powered cells that were the forerunners of animal life on Earth.

Of course, the question when the first living organism arose depends on the definition of life. A virus particle, for instance, is a collection of nonliving chemicals when it is outside the cell. But if life is defined as the ability to utilize chemical building blocks for an organism's reproduction, then once the virus particle enters a cell of a higher organism, it springs to life, whether that organism is a bacterium or an elephant. Most scientists now define life as a self-reproducing system that could have consisted originally of a few molecules replicating themselves.

What's clear, though, is that within a little more than a billion years of the Earth's formation, microscopic, single-celled organisms resembling bacteria were already in existence. Discoveries of microfossils in rocks from South Africa, Australia, and elsewhere testify to that fact.

But the next leap, to cells with nuclei, or eukaryotic cells, and to multicellular organisms, took twice as long, or two billion years. This

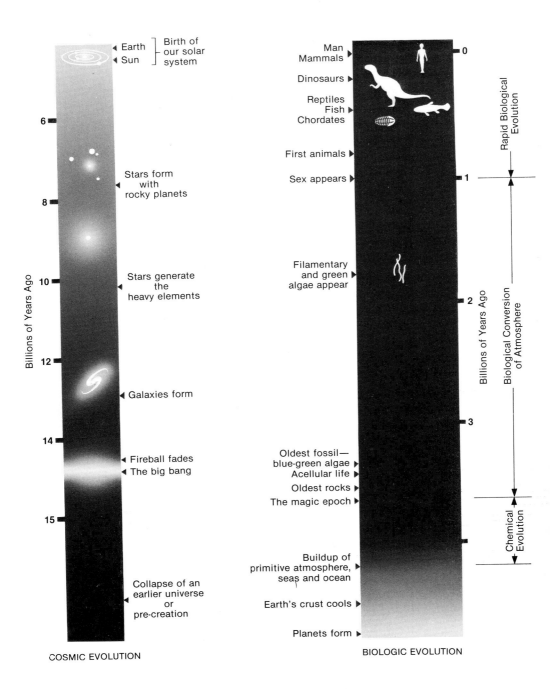

The long march to man.

has led scientists to conclude that the transition from nonbiological organic, or carbon-containing, matter to simple life was easier than the step from one-celled bacteria to multicelled organisms.

To be sure, this dramatic idea might be just an illusion, because the history of the first billion years of life on Earth has been almost completely erased by geological upheavals. Besides, the earliest life-forms probably were soft-celled bacteria, which would not have left a fossil record. From the first cell came many-celled creatures; then came the crustaceans; next the fishes, which were the first animals with a backbone, and eventually man. Our earliest ancestor, though, was not a microbe or a plant-like cell releasing oxygen via photosynthesis, but a different type of cell, which had learned to utilize oxygen instead of sunlight for energy production.

The big question is whether a sequence similar to the biological progression shown in our illustration could have taken place on other planets, and whether it should have led to similar results. First, it's probably the height of egotism to say that our solar system is unique in the universe and that no other such system exists. Logic cries out against such an assumption, considering what we now know about the universe.

But science demands proof. It will be some years still, possibly another decade, before planets outside the solar system are unequivocally detected. (See Chapter Nine, "Planet Search.") Measurement of their atmospheric composition as well as other factors would indicate whether an environment suitable for life exists on those planets.

We already have highly encouraging hints, however, that planets should be plentiful. About half of the stars visible through Earth-based telescopes are parts of multiple systems. Smaller suns circle bigger suns in close proximity. (In our solar system, Jupiter is big enough to have been a small star if its internal temperatures had been somewhat higher.) The presence of so many multiple-star systems suggests that gas clouds frequently break up into separate clouds of varying sizes, some forming stars and others forming planets.

Stars are divided into a number of categories, and only some can have planets with conditions suitable for life. The most famous classification of stars is the Hertzsprung-Russell Diagram (seen on the next page), in which stars are plotted with their luminosities and absolute magnitudes, or temperatures, as coordinates.

The diagram consists of a "main sequence," on which stars spend most of their lives, a giant branch, and the white-dwarf sequence.

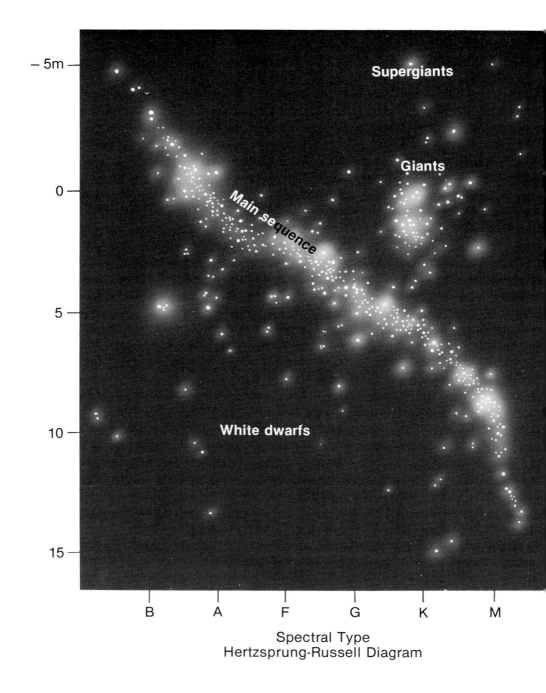

Spectral Type
Hertzsprung-Russell Diagram

Stars on the main sequence range from minuscule red dwarfs to blue and red giants, some of which are nearly as big as our whole solar system and forty times the mass of the sun. These giant stars burn their fuel too fast for any of their planets to develop life—the big stars are gone from the main sequence in a few million years.

The dwarfs, on the other hand, stay on the main sequence for billions of years, yet many of them are not really suitable to warm up life-bearing planets. To start with, many of the small red dwarfs flare up in a variable fashion—which would be lethal to any nearby life. What's more, the smallness of the dwarfs would make the life zone extremely narrow, so narrow that a planet close to the red dwarf might become forever locked in a position where it always presents the same face to its sun or is alternately baked by its sun and plunged into the numbing cold of space, the way Mercury is.

That leaves sun-like stars as the best candidates.

Such stars could range in mass from a bit greater than the sun down to about two thirds of it, with brightness from about twice as great as the sun to about one fifth. The life expectancy of such stars ranges from 10 billion years for our sun to 30 billion for the smaller stars. If long-lived stars have inhabited planets, life there would be unimaginably more advanced than ours.

As for the question of how many stars have planets, the answer, even when confined strictly to sun-like stars, is astounding. Our sun is a yellow dwarf of spectral type G. Of the 100 billion stars in our galaxy, type-G stars account for about 8 percent, or 8 billion. Scientists estimate that perhaps one fifth of G-type stars have planets—1.6 *billion* stars with planets in our galaxy alone. And there are 100 *billion* detectable galaxies in the universe.

But to be eligible as an arena for the emergence of life, a planet must be situated in a rather narrow zone of life, a kind of oasis between the Sahara and the Antarctica of space—the blazing heat close to the sun and the frigid cold of the solar system's outer reaches.

In our solar system, this zone encompasses Mars at its outer border, while Venus is close to the inner limit of the zone. The Earth appears to be ideally situated, in the middle of the zone.

Even if a planet is situated in the life belt, there is no guarantee that life on it will emerge. We now know, thanks to investigations with spacecraft, that both Venus and Mars harbor no readily detectable life, although some scientists continue to argue that life might have evolved on Mars in the past and might exist there even now in the form of microorganisms.

Astrophysicist Robert Jastrow, founder and director of NASA's Goddard Institute for Space Studies, as well as professor at both Columbia and Dartmouth, has argued forcefully, for instance, that data gathered by the Viking spacecraft that were landed on Mars in 1976 show evidence for the existence of microorganisms.

The Martian biology experiments were tests for the kind of life we know on Earth. The first experiment probed for gases given off by living organisms: oxygen by plants (or plant-like bacteria) and carbon dioxide by animals (or advanced microorganisms). When a Martian soil sample was moistened by a nutrient broth, a large amount of oxygen was produced. But the release of oxygen was immediate and sharp, strongly suggesting a chemical reaction.

The second test was more encouraging. It showed a continuous release of carbon dioxide, suggesting that some kind of microbe was metabolizing the food. The carbon in the nutrient was made radioactive for easier detection, and release of radioactive carbon dioxide made the Geiger counter aboard the spacecraft click thousands of times.

But this reaction, too, could be explained as a chemical one. So the experiments were repeated forty-six hundred miles away, the landing site of the second Viking. This time, the soil sample was dug up from under a rock. The sample thus had been shielded from ultraviolet radiation from the sun and, accordingly, should have contained fewer oxides. The first test, indeed, produced only about one tenth as much oxygen as did the sample at the first landing site. But it gave off just as strong a "microbe" signal as did the first sample.

Heating the soil sample to 110° Fahrenheit (43° C) left the oxygen release unaffected but depressed the "microbe" signal, suggesting the presence of microorganisms, since some of them would have been incapacitated by the heat. Finally, soil samples were stored aboard the spacecraft, and the tests were repeated some weeks later. The chemical results were unaffected, but the "microbe" signal was now gone. Without food, the "microbes" apparently had died.

Jastrow argues that all these fascinating data put the chance of life on Mars between 70 and 80 percent. Even though no carbon-containing molecules, which would have indicated the presence of microorganisms directly, were detected in the Martian soil samples, that wouldn't preclude the existence of Martian microbes. The test for organic molecules was not very sensitive; if Mars has as many microbes in its soil as does Antarctica, for instance, the microbes would not have been detected.

Jastrow has support from a number of scientists, including Gilbert Levin, principal Viking investigator in the microbe experiment. Most other scientists associated with the Viking probes, however, are more cautious—too cautious, Jastrow thinks. They feel that chemical explanation of the results is more tenable.

"At the same time," concedes Richard S. Young, head of the space agency's planetary biology program, "many scientists think that the results do not rule out the possibility of life on Mars. Other sites that have more water should be searched for and studied before an unequivocal answer is given to a question of this kind."

It's possible, of course, that rivers once flowed on Mars—the now dry channels photographed by orbiting spacecraft look very much like riverbeds. With water flowing freely, life would have had a better chance to emerge.

The case of Venus, once believed to be Earth's "sister planet," is an even stranger one. Venus is closer to the sun but not much, a little over one fourth the distance. Venus is almost the same size as the Earth, being less than one fifth smaller. And unlike Mars, Venus has a heavy atmosphere.

On the face of it, Venus would seem to be a prime candidate for life in the solar system, and for a long time scientists and science fiction writers alike speculated what Venus might be like. It was described as both a jungle world and a planet covered with an ocean of oil or carbonated water.

It was not until the space probes cast aside the veil of mystery that Venus was revealed as a celestial hell, with temperatures high enough to melt lead and with a corrosive atmosphere of sulfuric-acid droplets—hardly an inviting setting for the development of life.

Mercury, the planet closest to the sun, is even less inviting. It has no air or water, generally resembles the moon in its pockmarked appearance, and is mercilessly baked by the sun.

At the other extreme, Pluto, the outermost planet, is a frozen world. The four gas giants—Jupiter, Saturn, Uranus, and Neptune—contain the basic molecules needed for life and may even have rudimentary life-forms, but they lack conditions for complex life to develop.

Discovery of even the simplest kind of life on another planet—or remnants of life—of course would remove the creation of life from the realm of the miraculous and tell us that life is an inevitable outcome of chemical evolution of matter everywhere.

There is already strong suggestive evidence that that is indeed the case.

This evidence comes from discoveries by radio astronomers within the past decade that life's basic components are even now being manufactured in interstellar space.

It has long been known that many of the faraway stars contain the same elements as the sun, but the detection of complex molecules in space, which are precursors of life itself, came as a big surprise. Scientists had thought that starlight had enough energy to break apart any complex molecules that might form in space. No one had realized that collections of such molecules could be shielded by clouds of interstellar dust.

Among the molecules detected in dense molecular clouds are such familiar substances as water, carbon monoxide, methane, and ethyl alcohol. There are also precursors of the simple amino acid glycine, as well as more complex molecules. Scientists would not be surprised soon to discover simple amino acids in space—the building blocks of proteins.

Understandably, many scientists have exulted over the findings that the stuff of life is everywhere. "Anywhere you direct your radio telescopes—what do you see?" asks Dr. Ponnamperuma. "You see the very things we get in our laboratory mixtures.

"You look at the interstellar molecules and you see cyanide and formaldehyde. These two can provide the pathway for everything else.

"There is a simplicity in the whole scheme—so much so that you practically feel that the whole universe is trying to make life!"

These startling discoveries have led some scientists to revive in a new guise the old theory of "panspermia," the idea that the seeds of life drifted to Earth from elsewhere across the gulf of interstellar space. In this new version, though, the seeds of life do not originate on another planet, as they did in the original theory.

The authors of this idea, called appropriately "the life cloud," are the British astrophysicist Fred Hoyle and his collaborator Chandra Wickramasinghe. In their scheme, clumps of grains composed of organic molecules needed for life arise in prestellar molecular clouds, where stars form. As the stars are "turned on," the grains are kicked out and disperse through interstellar space. Some of these grains, according to this theory, reached the Earth billions of years ago as meteorites containing the molecules needed for life to arise.

But it isn't really necessary to invoke a panspermia-type hypothesis for the origin of life on Earth—as all those experiments in the laboratories demonstrate.

With that incredible life machine at work in the universe, creating the constituent molecules of life, which radio astronomers have discovered in recent years, life would have arisen independently in all the nooks and crannies of the cosmos. With the mind-numbing multitude of appropriate settings on hand, the diversity of life-forms in the cosmos is bound to be nothing short of astounding. On Earth alone, as we shall see in the next chapter, the diversity of life has been far greater than we realize, hinting at directions life can take on other planets.

CHAPTER THREE

Conquest of the Land

Another billion years passed from the formation of the first cells that used oxygen for energy to the emergence of the first multicellular organisms. The seas were becoming filled with increasingly more complex oxygen-breathing multicelled forms of life: lobster-like trilobites, sea worms, jellyfish, shellfish, sponges. Primitive plants dotted the ocean floor.

The ocean teemed with life, but the great world continent, Pangaea, and the few smaller landmasses, were still barren.

Then, 400 million years ago, onto the shores of the supercontinent gradually drifted blue-green algae, the most primitive of plants. They established the first foothold on the continent by struggling for a tenuous existence on mud flats along rivers and streams.

Seaweeds floated ashore to join the algae. Life had left the sheltered environment of the sea. The invasion of the land was on, opening up undreamed-of evolutionary opportunities.

Nearly 3 billion years had elapsed between the time the first living molecules appeared and the beginning of the conquest of the land.

Why did it take so long?

For one thing, the searing heat that enveloped the early Earth began to decline from around 160 degrees Fahrenheit (71 degrees C), to 126 degrees Fahrenheit (52 degrees C), only about 1.3 billion years ago. High temperatures on the early Earth precluded the development of more-complicated cells than simple bacteria. It took a billion years for the ocean to cool sufficiently and to become oxidized to allow the emergence of the much-more-complex cells that have a nucleus, rather than of the primitive cells with a single, looped chromosome.

That lower temperature is just below the destruction level of complex proteins, so that now the complicated cells could survive.

For 2.7 billion years the simple anaerobic bacteria, those that thrive in the absence of oxygen, were the only life on Earth.

Now, with the atmosphere acquiring its first oxygen thanks to the evolution of the photosynthetic apparatus, the eukaryotic cells (those more complex cells, with a nucleus), were better equipped to deal with the presence of free oxygen than were the bacteria. Instead of the fermentation of glucose, the bacteria's way of getting energy, some eukaryotic cells 600 million years ago evolved a metabolic pathway that enabled them to use oxygen for energy—increasing their energy production eighteenfold.

This giant leap in energy yield fueled the development of higher animals, including trilobites and other slow-moving creatures such as brachiopods.

This happened during the Cambrian Period, which lasted from 570 million to 500 million years ago. It was followed by the Ordovician, a time of major expansion of the invertebrates: marine animals with mineralized skeletons and shells; cephalopods, or mollusks such as squid and octopus; as well as starfish and corals. There were also the forerunners of fish.

These creatures stayed in the sea until the oxygen level in the atmosphere reached sufficient concentration to serve as an efficient shield against deadly radiation from the sun.

About 400 million years ago, the ozone shield was finally in place, and with shallow seas covering large areas of Pangaea, primitive plants began to establish a beachhead on the fringes of the land.

To cope with the new environment, the primitive plants had to evolve a number of advanced features.

They developed a protective layer called epidermis punctured at intervals by pores called stomata, a central strand of tissue called the xylem, and phloem tissue, which transports sugars and other substances made by plants.

The primitive plants also displayed terminal or lateral organs containing reproductive spores, cells with a thick protective coat.

Plants evolved leaves, which gave them a great capacity for food manufacture. Gradually, plants developed the ability to increase the diameter of their stems almost indefinitely. Land plants could now grow to tree size, and forests began to spread through Pangaea's lowlands.

At first, these were small trees, about three feet tall. Later, though, conifers evolved a big improvement on the spore: the seed. It became the plants' key to life on dry land. Once these breakthroughs occurred, the vast continent was open to expansion by the plants. Huge tropical forests began to cover the land.

Almost in parallel with the plants, animals began to hit the beaches of Pangaea. Plants lured them ashore; now there was food on the once-barren land. The spread of forests afforded an arena for further development of the animal newcomers.

The first animal colonists were most likely the scorpions, members of the arthropod phylum, which also includes crustaceans and insects. Equipped with legs and hard, external skeletons, the arthropods were already partially adapted to life on land.

Meanwhile, in Pangaea's extensive shallow ponds darted the first freshwater fishes.

Two subclasses of fishes apparently arose independently at about the same time to cope with two conditions of aquatic life. The bony fishes evolved in freshwater streams, where they had to cope with fast currents as they headed upstream, a trait freshwater fishes retain to this day. This called for a strong backbone; these fishes were the first vertebrates.

The second subclass consisted of cartilage fishes, which lived in the ocean. Some, such as the lamprey and the shark, still survive. Most fishes living today, though, are bony fishes, the class that turned out to be more successful and came to dominate both streams and seas. In the seas, their speed allowed them to stay out of the clutches of the huge aquatic scorpions known as eurypterids.

The ancestor of the land vertebrates was a fish with two pairs of fins. Fish originally had a pair of continuous fins, one on each side of the body, serving as bilge keels. The middle part of these fins disappeared, because typical wave-like body movements emphasized the development of the front and back parts.

The two pairs of fins are essential for controlling the up-and-down plane, keeping on a level keel, initiating rolling and yawing and other movements—much as an airplane has to control its flight.

Nature emphasizes simplicity, so the two pairs of fins are an efficient means of propulsion in water.

Similarly, on land, the fewer feet the more efficient it is for a large animal to walk and run. Even land creatures that originated as worm-like, centipede types that originally had a large number of small legs, reduced the number of legs as they grew larger. The mechanics of getting around demanded fewer, bigger, and longer legs to lift the body and lengthen the stride. Insects, derived from the centipede-millipede type of creatures, settled on three pairs of legs. Spiders, which came from the aquatic horseshoe-type crabs, settled for four pairs.

Several times in their early past, freshwater fish went down to the sea and gave rise to the bony fishes now there. Others stayed put as the freshwater bony fishes of today.

During the Devonian-Carboniferous period, freshwater areas became shallow and oxygen-short. The bony fishes evolved lungs; in those that returned to the sea, the lungs became the hydrostatic air bladder.

The fishes that stayed in the freshwater areas began to push along the shallow bottoms of ponds, from one wet place to another, gulping air on the way. Inevitably, these fish got better and better at breathing air and using their fins for shoving their bodies along. Particularly skillful at trundling from shallow pool to shallow pool were the crossopterygians; the structure of their fin bones made for fairly easy locomotion on land. "They didn't 'choose' to go onto the land," says N. John Berrill, the noted English-born biologist. "They suddenly found themselves there and made the best of it."

The fish gradually evolved into amphibians resembling short-tailed alligators. Such amphibians lay in sluggish streams, awaiting their prey. These animals spent at least part of their adult life on land, where they breathed air. But they retained their link to the sea, for they still laid their soft, fish-like eggs in water, and their larvae had to live in water. The word amphibian in Greek means "double life" (from *amphi*—double, and *bios*—life), referring to the two phases of the amphibian's existence.

The early amphibians gradually evolved into a number of shapes, ranging from the snake to the short-bodied, four-legged turtle. Frogs are descendants of early amphibians and have existed in the familiar form we know today for the past 180 million years. Frogs, of course, continue the amphibian "double life" to this day.

The big and final break with the sea for most amphibians then living came later, in the Pennsylvanian (Carboniferous) age, 275 million years ago. This was the change from the four-legged amphibian to the reptile. It involved development of stronger legs so as to lift the body off the ground, better hearts and lungs, and most important of all, the evolution of the hard-shelled egg, a self-contained capsule complete with food available as yolk and albumin as a substitute for water.

Unlike the amphibian soft eggs, the reptilian eggs were fertilized inside the female's body. The embryo, developing inside the egg, was surrounded by the amnion, a fluid-filled sac; hence the name amniote egg. Inside the amniotic sac, a partition contained yolk for food; another partition served as a waste-disposal sac. The whole was wrapped in a tough shell.

Human reproduction also is based on the amniote-egg principle, illustrating man's direct link to the reptile. The mammalian system, of course, differs from the reptilian, but there are basic similarities. The human egg is also fertilized inside the mother's body and then becomes surrounded by an amniotic-fluid-filled sac much like that in the reptilian egg. This salty fluid is the reminder of the sea from where those ancestors of both men and reptiles—the fishes—came. The human egg covering even contains a tiny yolk sac, even though the embryo is supplied with nourishment through the placenta from the mother's bloodstream. The wastes are also withdrawn through the placenta, which may be a restructured waste-disposal sac of the reptilian egg.

In any event, with the advent of the amniote egg, reptiles were free to venture into the interior of the continent, since they no longer depended on water for reproduction. The reptiles took the ocean with them onto the land and repackaged it into an egg. It was their counterpart of the plant seed. A miniature adult could now develop and step full-fledged onto the dry land instead of darting out as a little aquatic larva.

As the climate gradually changed to a desert-like one, about 200 million years ago, the first dinosaurs appeared. This was also a time of great geological upheavals; many different, self-contained environments were created on Earth that became populated by strikingly differing as well as strikingly similar animals. The fact that similarities evolved under differing circumstances, and that contrasts developed within similar circumstances, hints at the rich possibilities for life to take on its paths of evolution.

In the next chapter, we shall look at what other turns the evolution of intelligent life on Earth could have taken—and could take on other planets.

Part II

DARWIN'S UNIVERSE

CHAPTER FOUR

The Turns and Twists of Evolution on Earth

In the mists of prehistory, you might say three planets named Earth circled the sun. On each Earthly miniplanet, furthermore, evolved different forms of life, hinting at the directions life can take elsewhere. The story of the three planets named Earth may sound as if it comes from the wild imaginings of an Immanuel Velikovsky. It doesn't. The existence of the three miniplanets named Earth is, instead, a historical and scientific reality stranger than fiction.

The three Earths came about because, 200 million years ago, giving way to stresses and strains deep in the interior of the planet, the great world continent, Pangaea—the original "Earth"—unceremoniously split, as shown in our illustration. To the north lay Laurasia, containing what later became North America, Greenland, Europe, and Asia. To the south, the big landmass was Gondwanaland, comprising what later became South America, Africa, Australia, and Antarctica.

The breakup of Pangaea didn't stop there. Later still, smaller fragments began to leave the two main continents. First, about 180 million years ago, Australia and Antarctica broke off in one piece from Gondwanaland and drifted south. Then, about 130 million years ago, North and South America began to separate from their respective mother continents as shown in our illustration on page 37.

These continents in motion were not exactly like Noah's arks, however, because they were not staffed with two animals of each species. In fact, the combined continent of Australia and Antarctica had sailed away like a big raft without any mammals on board—only insects, dinosaurs, and those strange half reptiles, half mammals known as monotremes, of which the platypus is a representative that survives to this day. The first mammals did not appear in Laurasia and Gondwanaland until about 25 million years after the departure of Australia and Antarctica.

Meanwhile, Laurasia and Gondwanaland had drifted together once again; thus mammals were able to spread across both those landmasses. So when North and South America began to break away, they already had mammals, as well as dinosaurs. The separation of North and South America was a convoluted dance. North America had never really separated fully from Laurasia, to which it remained connected by the land bridge that existed until about forty thousand years ago where the Bering Strait is now. South America, in turn, first became attached to North America at the Panamanian Isthmus but later broke that link, only to be reunited again about 10 millions years ago.

Across those land bridges, in the meantime, poured animal hordes. The traffic was especially busy across the Bering bridge, with animals migrating in both directions. The horse and the camel, for instance, originated in North America and then spread to Europe and Asia.

Despite the Bering bridge, though, animals in North and South America and in Laurasia-Gondwanaland were beginning to develop in somewhat different directions. These differences became much more pronounced as South America separated from its northern neighbor, becoming a separate continent like Australia-Antarctica.

So when all this bumping together, parting, and alternate creation and submergence of land bridges slowed down, continental drift had created three isolated miniplanets on Earth, three theaters with different casts of animal actors set to play out the drama of evolution in different settings and under different conditions. There was the North American-European-African-Asian arena as one miniplanet, the largest of the three; and there were South America and Australia-Antarctica as the other miniplanets.

By this time, tens of millions of years had passed from the original breakup of Pangaea. Already living in semi-isolation for a long time, many animals began to develop so differently from their counter-

The supercontinent Pangaea.

Pangaea's breakup.

The world today.

parts elsewhere that they might as well have been living on different worlds. Their remotest common ancestors were mere blobs of living matter, not much like the end products at all. Now these animals were ready to develop in still newer directions.

It was as if three playwrights had conceived three plays on the same theme and three designers had created the sets. The curtains rose almost simultaneously in the three theaters.

What happened on the stages of those miniplanets in terms of both evolutionary diversity and similarity of animal species is what is likely to happen on life-bearing planets elsewhere. As we shall see in the next chapter, the laws of life appear to be as uniform throughout the universe as are the universal laws of chemistry and physics—on which all life is based. In terms of life, it's Darwin's universe, with significant but not outlandish local variations on the theme with which we are familiar.

Each of those miniplanets on Earth, therefore, could be a mirror of another planet elsewhere in terms of life, while our oceans represent still another distinct setting for another type of life to evolve. (See Chapter Six, "Ocean Kingdoms and Insect Worlds.")

By scrutinizing the history of those miniplanets on Earth, we can get an inkling of how life will have evolved on other worlds. But before settling back to see that drama unfold in three theaters on Earth, before picking up the thread of our story again, to understand more clearly what happened on our miniplanets and what other evolutionary paths life can take elsewhere, we must look briefly at Charles Darwin's towering contribution to our understanding of life on Earth.

For Darwin, this understanding began with a study of a smaller-scale miniplanet, the Galápagos Islands—a study that helped him formulate his theory of evolution more than 120 years ago.

While journeying around the world for five years as an unpaid naturalist aboard the British survey vessel H.M.S. *Beagle*, the young Darwin became intrigued with the problem of the origin and diversity of species.

In the Galápagos Islands, which lie about six hundred miles west of South America, Darwin noticed that each island had its own form of tortoise, of mockingbird, and of finch. The various forms were clearly related, yet they differed from one another.

Darwin's study of the Galápagos finches has become a classic, but he actually started out with the sea turtles.

He discovered that the tortoises on one island were distinct from those on another island, and he couldn't accept the idea that they had been separately created in the two islands, or that one kind had gotten to one island from the mainland, another kind to another island. Darwin concluded that the first migrant tortoises had simply spread among the islands.

On each island, they were not in competition with those on other islands, and they became gradually different, going off into slightly different ways of living, and they were recognizably different. Tortoise hunters Darwin met told him they could tell what island the tortoises had come from.

Darwin's flash of insight was the realization that all the Galápagos tortoises stemmed from one species, and he immediately applied it to the finches. He found that instead of having regular finches as we know them, fourteen varieties of finch populated the islands. Feeding on different kinds of seeds and vegetation, they had developed different types of beaks, with one finch even imitating the woodpecker by using a castus spine as a tool to dig insects out from under the bark.

Wrote Darwin in *The Voyage of the Beagle:*

The natural history of these islands is eminently curious, and well deserves attention. Most of the organic productions are aboriginal creations, found nowhere else: there is even a difference between the inhabitants of different islands; yet all show a marked relationship with those of America, though separated from that continent by an open space of ocean, between five and six hundred miles in width. The archipelago is a little world within itself, or rather a satellite attached to America, whence it has derived a few stray colonists and has received the general character of its indigenous populations. Considering the small size of these islands, we feel the more astonished at the number of their aboriginal beings, and at their confined range. Seeing every height crowned with its crater, and the boundaries of most of the lava streams still distinct, we are led to believe that within a period geologically recent the unbroken ocean was here spread out. Hence, both in space and time, we seem to be brought somewhat near to the great fact—that mystery of mysteries—the first appearance of new beings on earth.

On his return to England, Darwin continued to think about his observations. His conclusion that the same species had reached the islands and then evolved into slightly different ones was reinforced by the long experience of plant and animal breeders, which clearly showed that artificial selection could produce evolution by breeding. For more than twenty years, Darwin quietly considered this line of reasoning. Finally, in 1858, he offered his explosive explanation for the evolutionary changes: natural selection. The following year, Darwin published *On the Origin of Species by Means of Natural Selection*.

The theory of evolution proposed by Darwin suggested that all living beings belong to one great family and that differing species evolved from this family through natural selection, or adaptation to their environment.

Among Darwin's postulates were the ideas that species change continually, with new ones emerging and others becoming extinct; and that the process of evolution is gradual and continuous. He saw natural selection as being the key to the evolutionary scheme. He defined natural selection as a two-step process. First is the production of heritable variations through favorable mutations, or changes in the creature's genes, as well as through a recombination of genetic material. Second is selection through survival in the struggle for existence. Selection led to better-adapted individuals through a gradual transformation of their populations.

Darwin had said that natural selection is not a matter of pure chance, since the random variations are sorted out by selection for survival—an "anti-chance" factor. But Darwin did not completely exclude the element of chance in evolutionary development.

In keeping with that idea, Darwin suggested that, rather than proceeding with a straight-line directiveness, evolution flourishes like a living, supersensitive tree, sending out intricate, ever-changing branches into the environment. There is nothing preordained about the success of the branches. In fact, most of them wither and die.

The tree-branch analogy is typified by the evolution of the horse. Textbooks tend to illustrate equine evolution by showing pictures of progressively larger horses, starting with a cat-sized founder of the clan.

But that's not exactly how horses evolved. There were many offshoots of a horse-like animal, evolving in various places at various times. Some grew bigger, some got smaller—there was no arrow-straight gallop toward today's horse. What happened, rather, was an explosion of slightly differing types on one branch of the evolutionary tree. Of these, one gave seed that grew into today's horse. In a crowd of similar animals, only one survived.

Elaborating on this idea, George Gaylord Simpson, the noted American paleontologist, likens evolution to a lottery in which most of the ticket holders are losers—they become extinct—while some receive small prizes and a few become big winners. "Man simply happens to be the descendant of a long line of organisms that drew winning tickets in every successive adaptive radiation," he writes in his book *Life*. "The basic adaptations of his ancestors have proved, in hindsight, not to have closed out the evolutionary future." (Adaptive radiation is the spread of animals into new ways of life, or ecological niches.)

Simpson has emphasized that evolution is recklessly opportunistic. Evolution awards the survival prize to a species in which any variation yields a competitive advantage either over other members of the organism's own population or over individuals of a different species.

This is the process that has fueled evolutionary progress ever since life began. It's a process not guided by a prescribed program but the result of instantaneous, spur-of-the-moment decisions of natural selection in the context of constant interplay of environmental and genetic influences.

In this dynamic scheme, dinosaurs, although they ruled the Earth for almost 100 million years, obviously were losers. Unable to adjust to whatever drastic environmental changes they faced, they began to die out 75 million years ago, and within about 10 million years—an instant in geological time—the giant reptiles had vanished.

When dinosaurs ruled the Earth, in the background lurked the first mammals, tiny, shrew-like creatures that themselves had stemmed from a type of reptile.

Once the dinosaurs were gone, mammals and such descendants of the dinosaurs as birds surged into the niches vacated by the giant reptiles. It was a striking example of adaptive radiation. Where the sinister-looking flying reptiles, the pterosaurs, once soared, now birds filled the skies. Where the reptilian ichthyosaurs once leaped in the ocean waters catching fish, now dolphins frolicked—mammals that went back to the sea. Where tyrannosaurus once trampled grass, now rhinoceros-like beasts roamed.

Geographically, as the Age of Mammals began, those three continental theaters were set to play the same play with different casts of characters.

The original cast consisted of small, primitive placental mammals and small, primitive marsupials, or pouched mammals, which give birth to premature babies, which then grow up in their mothers' pouches.

In South America, which by that time had become an island continent separated from North America, there was also a strange group of mammals known as xenarthrans. Although part of the placental variety of mammals, these creatures included such unusual beings as the armadillo, the sloth, and the anteater—all rather closely related to each other.

In the main theater, consisting of Europe, Africa, Asia, and North America, united at the Bering bridge, marsupials lost out almost entirely, and placentals took over the stage, with only the opossum surviving in North America as the sole representative of marsupials.

Marsupials flourished in South America, however, alongside some placentals. A tip of Antarctica then adjoined South America. An opossum-like marsupial apparently got to Antarctica and from there wandered all the way to Australia.

Thus, the three theaters were now set up: Europe-Africa-Asia-North America as one miniplanet, with an exclusively—except for that North American opossum—placental cast; South America, a miniplanet with a mixed cast of primitive placentals, marsupials, and even some small prosimians, primates that apparently had drifted across the ocean from Africa; and Australia, a miniplanet with a marsupial immigrant and a small indigenous population of those odd creatures known as monotremes, whose females both lay eggs and feed their young with milk, as the platypus females do.

The larger-scale events unfolded in the African-Asian-European-North American arena. The primitive placental stock diversified into the mammalian kingdom as we know it, with whales, bats, and man eventually evolving from one subgroup of shrew-like creatures.

On this largest of landmasses, generally evolved the largest and most intelligent animals. On the smaller landmasses, on the other hand, evolved somewhat smaller beasts, or at least the smaller ones survived for longer periods of time. In fact, there is a striking relationship between the area of an island and the number of animal species found on it; the number increases with the size of the island. Not only that, but seem to evolve larger in size on bigger landmasses, too, as our illustration indicates.

Australia, on the other hand, offers a somewhat reduced example of evolution but an equally fascinating one for all that.

The primitive pouched opossum that reached Australia showed how a single creature, given enough time and a big enough stage on which to act, can give rise to a veritable zoo. It's an example of extreme diversification of a solitary animal that evolved in all the directions allowed by its own genetic potential and the environment it encountered.

From that single marsupial ancestor, developed the kangaroo, the koala, the phalanger (which resembles a flying squirrel), a native mouse, a cat, a mole, a sloth, the groundhog-like wombat, a pouched "wolf," and even an anteater. These creatures climbed trees like monkeys, soared through the air, burrowed into the ground. Some became carnivores, others vegetarians. Some were tiny, others big. All possible variations were explored. Although all these animals, except the kangaroo and the koala, resemble their placental counterparts on other continents, they are only faintly related to them.

Elsewhere, the placental mammals—the "true" mammals—radiated to fill the ecological niches available to them. In Australia, the marsupials did it because there was no competition. Australia had only some rats and bats as representatives of the placental mammalian legions, and these didn't come until much later, probably island-hopping from Indonesia.

Isolated from the rest of the world, Australia and its outlying islands emerged as biologically the most interesting region in the world. What happened in Australia could happen on another planet on a larger scale.

The Australian experience provides sharper glimpses into the basic processes of evolution than that of any other continent.

Isolation reduced outside interference to a minimum, so that the marsupials and those descendants of the dinosaurs, birds, could safely spread into all the varied niches. Encouraged by the absence of predators and by the vacancies in environmental settings, the creatures diversified into ways of life that were not exploited elsewhere. In New Zealand, for instance, there are only two "old" mammals, both of them bats. So New Zealand's birds filled many mammalian niches.

Free from predators, some types of birds both in Australia and in New Zealand, abandoned flight as a way of life, with the kiwi and the woodhen, for instance, choosing life on the ground instead.

ZEBRA   GORILLA   RHINOCEROS

KANGAROO   SHORT-NOSED   PLATYPUS
                 BANDICOOT

AFRICA   ELEPHANT   GIRAFFE

Large animals emerge on large continents.

AUSTRALIA   WOMBAT   KOALA

Smaller animals inhabit smaller continents.

The most famous of these flightless birds was the moa, of New Zealand. It stood twelve feet tall on legs as thick as those of a steer. Darwin was the first to suggest that the moas evolved to fill a niche left empty by the absence of mammals. In the thick forests of New Zealand, the moa evolved in another direction: into the small kiwis, which hide in underground burrows and emerge only at night to feed. In Australia, the moa evolved into the five-foot-tall emu, which still survives. It can run as fast as forty miles an hour and is the bane of farmers and ranchers, because it is strong enough to crash through paddock fences.

The Australian region's experience shows how in a protected setting a diversity of animal and plant life can arise, thanks to the opportunities provided by the environment to match any such diversity that has been created on other continents through intermingling, invasions, and various upheavals. Due to its isolation, trends in the world at large left Australia largely untouched. To cite another example, there still lingers a lunged cousin of air-breathing fishes in the rivers of northeastern Australia.

In all, there are one hundred thousand species of plants and animals in the Australian region, a high number in proportion to the size of the land.

Australia's even-tempered climate in the past no doubt contributed to this richness of living things.

Equally fascinating is Australia's and New Zealand's plant life. Australia's slow-motion journey north—the continent is believed to have moved three thousand miles in the past 100 million years—wrought dramatic changes in the vegetation. In the beginning, climate changed from temperate to arid. Fossil evidence reveals that rain forests of beech trees once covered most of southern Australia. Great rivers cut through the forests. Now only dry riverbeds remain, and grass and scrub cover the land—evoking pictures of those dry riverbeds photographed on Mars.

Acacias and eucalypts replaced the beech forests. The eucalypts diversified into more than six hundred species, the largest growing as tall as three hundred feet, with a trunk six to nine feet in diameter, second only to California's giant sequoias.

Gradually, a lot of the lush forests vanished as the southern part of the continent dried up. Yet it is believed that Australia always had pockets of aridity, explaining why much of the existing fauna and flora are so well adapted to arid conditions.

Rain forests and dense eucalyptus forests still cover parts of Australia. Today's Australian trees are different too. Some species have male and female trees. The grass tree displays mops of Harpo Marx-like hair and white flowers on long spikes. The bottle tree looks like a bottle, and the wattle produces pods like peas.

Australian flowers are just as unusual, and strikingly beautiful. There's the kangaroo paw, which is furry to the touch; the waratah is a shrub that grows flowers the size of a grapefruit. Dazzling orchids grow through much of Australia.

Similarly to Australia, New Zealand parted company from the main southern landmass about 160 million years ago. Only ferns and tuataras, reptiles more primitive than lizards or dinosaurs, were on board this incompletely staffed life raft. The tuatara still survives on a few remote New Zealand islands, a strange creature with a relic of a third eye situated Cyclops-like in the center of its head. This "eye" is now hidden deep beneath the lizard's skin and apparently isn't functioning. Its original purpose probably was not to act as a true eye but as a light window that would warn the animal of approaching danger or announce the start of day or night. The remnant of such an eye is situated in our heads too. It has been restructured into the pineal gland.

Ancient ferns jam the islands' forests, and strange insects and birds flit about—reminding us just how "alien" life can be even on Earth.

South America evolved as a strikingly different miniplanet.

To start with, South America was connected to North America around the end of the Cretaceous and the beginning of the Paleocene, or slightly more than 60 million years ago. It then became disconnected and remained an island continent for nearly 50 million years. Then it became connected to North America again by the Panamanian Isthmus about 10 million years ago and has been so linked ever since.

While South America was isolated, primitive rodents and other creatures island-hopped there from North America on islands where Central America now links the two continents.

Into South America also poured at various times the lemur-like prosimians from Africa, marsupials from North America, and still later another flood of migrants from North America. South America thus became the original American melting pot.

All this resulted in a fascinating mixture of animals that even today includes such strange creatures as armadillos, capybaras (the largest rodents on Earth), llamas, tapirs, tree sloths, and many others.

The South American mammals that became extinct were stranger still. There was, for instance, a fierce saber-toothed tiger with a pouch like that of a kangaroo; a giant anteater as big as a horse; a tank-like, armored plytodont; a giant sloth, and many others.

George Gaylord Simpson has called South America "an almost ideal natural experiment," because we can follow evolution at work there for tens of millions of years.

When South America was separated from North America, during the Age of Mammals, the South American mammals evolved in almost complete isolation. Almost, but not completely, because two alien groups of mammals—primates and rodents—were introduced during the "experiment" to make it even more instructive.

The history of South America demonstrates the kind of evolution that could well take place on another planet with similar continents, where development of life has followed a course similar to that on Earth.

At the same time, the evolutionary development of South American animals differed from those in Australia. In Australia, all the marsupials came from a single descendant and proceeded to fill all the available niches. South American animals, on the other hand, stemmed from different roots. But all of them diversified without competing with each other, acting as if they, too, were all members of the same family.

Students of the South American animal world have learned that fossils of mammals there date back to early in the Age of Mammals, or about 60 million years ago. They have found a few scraps of fossils dating back to the late Cretaceous, which just preceded the Age of Mammals. These fossil fragments appear to have belonged to a primitive opossum that closely resembled its North American counterpart. Both of these creatures could have been the ancestors of the large variety of South American hoofed mammals and the numerous marsupials.

The next fossil find, dating back to 55 to 60 million years ago, revealed a richness of mammalian life that had diversified and expanded to an amazing degree. There were already three families of marsupials, five families of hoofed animals, and a family of those curious xenarthrans, then represented mostly by armadillos.

At the same time, all those mammals belonged only to those three basic groups: marsupials, xenarthrans, and hoofed herbivores. Such a spare mix of animals was unique. Except for Australia's, the continents' faunas each consisted of more than three stocks.

North America, for instance, had a much more varied animal population. At least seven orders of land mammals there were not derived from the ancestral marsupials, hoofed herbivores, or xenarthrans.

In South America, in contrast, the three basic stocks diversified in ways peculiar to their setting. In the mixed-up animal world of South America, a pouched marsupial carnivore might be chasing a placental herbivore. Even more strikingly, the low-slung, dog-like marsupial that looked like a combination of a dog and a cat, occupied the niche filled not only by dog- or wolf-like animals on other continents, but also that filled by cats elsewhere—another hint how animals on other planets might differ from ours.

Marsupials apparently arrived in South America from North America, but the Xenarthra appear to have evolved their distinctive features in South America, even if they did have ancestors elsewhere. The hoofed herbivores, or ungulates, such as the llamas, of South America, probably stem from the primitive order of Condylarthra, whose fossils have also been found in North America, Europe, and Asia.

The vestiges of the miniplanets can be seen on today's map of geographical distribution of animals in the world. Six major animal regions are discernible. First, there is the Paleoarctic, which includes Europe, the Soviet Union, North Africa, the Middle East, and most of China. These areas have in common such animals as the wild boar and fallow and roe deer.

On the other side of the globe stretches the Neoarctic realm, covering North America and Greenland. It has animal forms comparable to the Paleoarctic region plus the caribou, mountain goat, and muskrat.

Near the equator lies the Oriental realm, which includes the Indian elephant, Indian rhinoceros, macaque, gibbon, orangutan, and other beasts.

Straddling the equator are the Ethiopian and the Neotropical areas. The African elephant, lion, zebra, African rhino, ostrich, giraffe, chimpanzee, and gorilla are the inhabitants of the Ethiopian realm, while the armadillo, anteater, llama, tree sloth, toucan, and other unique creatures live in the Neotropical region, which includes South and Central America.

Finally, the Australian realm retains that marsupial zoo, with about

170 species, as well as monotremes such as the platypus and the spiny anteater.

The most striking fact about South America—and Australia, for that matter—is that many of the animals that evolved there mirrored the forms and ways of life of creatures on other continents, although the animals in the three arenas had come from different stocks and had developed in isolation, thousands of miles apart.

This mirroring, in fact, is so uncanny that for a long time it confused students of evolution.

Little wonder it did! Put side by side, the images of animals from the three miniplanets show strikingly similar features despite the fact that the continents have been separated for millions of years.

If a South American shrew-like marsupial had looked at the North American mirror, it would have seen a placental shrew; the South American marsupial "wolf" would have stared at a "real," placental wolf much like today's wolf; the slightly camel-like litoptern, which lacked a distinctive hump, would have gazed at a true camel; the tiny horse-like litoptern would have looked up at a horse; the toxodont would have looked down at the somewhat smaller rhinoceros; and the strange-looking ungulate homalodothere would have seen a bigger image of itself in the North American chalicothere, a creature that looked like a combination of a horse and a cow.

So great did the similarities appear that one early Argentine paleontologist got so carried away with the resemblances of the South American fossils to those elsewhere that he proclaimed that all mammals had originated in Argentina and from there had spread to the rest of the world.

What the Argentine paleontologist was seeing, however, was the fascinating phenomenon of evolutionary convergence, which was not yet recognized as such at the beginning of this century, when he proclaimed Argentina to be the birthplace of the mammals.

Convergence is the separate evolution of unrelated animals engaged in similar ways of life. It can lead to animal forms closely resembling their counterparts elsewhere. These near-mirror images may be not only geographically distant but separated by millennia as well. (Evolutionary parallelism, on the other hand, refers to evolution of *related* animals in different geographical settings.)

In terms of time separation, the ichthyosaur and the dolphin yield a convergent near-mirror image. The ichthyosaur is an extinct reptile that went back to the sea; the dolphin is a mammal that became what scientists call a "reentrant," about 20 million years ago, by also returning to the ancestral home of both reptiles and mammals.

Although the two animals are not related, the similarities of form between them are striking. The ichthyosaurs were reptiles, but their females bore living young as part of their adaptation to life in the sea. So do the dolphin females, of course.

There were some small differences, to be sure. The ichthyosaur had a somewhat longer snout than the dolphin, and its tail fin was vertical, while in dolphins it's horizontal. But the resemblances between the two are more pronounced than the differences.

Another intriguing example of convergence was the simultaneous emergence of the true wolf of Eurasia and North America and the marsupial, or pouched, wolves of South America and Australia, as shown in our illustrations on the next two pages. There is a good reason for such convergence, of course. All three carnivores evolved to chase other creatures, and the wolf-like shape fits that role the best.

The South American marsupial carnivores, not surprisingly, resembled the Australian marsupial wolf much more closely than they did the placental carnivores on other continents.

The two marsupial wolves are believed to have evolved separately, but they arose from similar primitive opossums. Accordingly, in the case of these two animals not only convergence but also parallelism is easily discernible.

In still another dramatic example, the hummingbird and the humming moth have only the remotest common ancestry, lost in the mists of time. For all practical purposes, their common ancestor was the amoeba, which obviously looks like neither a bird nor a moth. Yet they have converged so remarkably in physical appearance, in flying habits, and in feeding on the nectar of flowers, that at dusk and at a distance it is difficult to tell one from the other.

There are many more examples of convergent evolution: The porcupine, the hedgehog, and the echidna, for example, all evolved quills independently of each other and are only extremely distant relatives. The same is true of North American flying squirrels and the Australian flying opossums known as sugar gliders, or phalangers.

Convergence also occurs in organs of locomotion. To take to the air, for instance, four different creatures on Earth evolved wings: the reptilian pterodactyls, birds, bats, and insects. The first three evolved wings from walking legs, but insects grew them as original equipment.

Placental wolf.

Placental anteater.

Marsupial wolf.

Marsupial anteater.

Students of evolution have designated similar characteristics inherited from a common ancestor as being "homologous." Those that arose separately are called "homoplastic." In turn, if homoplastic features mark similar functions, they are called "analogous."

In this scheme, all bird wings are homologous, since they were derived from a common ancestral wing of the first bird, which stemmed from a type of reptile. Wings of bats and birds, on the other hand, developed at different times, are homoplastic and analogous. They are not homologous, even though they contain homologous bones originally derived from reptilian ancestors.

The insect wing, however, is merely analogous to the other three, because it arose independently in a completely unrelated fashion, not from walking legs but somehow directly from the body.

A surface convergence, though, can hide significant anatomical differences. A comparison of the dolphin with the shark—both have streamlined bodies—illustrates how animals that look somewhat alike can differ significantly in internal structure. The dolphin's surface resemblance (in shape) to a shark is obviously the result of the dolphin's adaptation to aquatic life. On closer inspection, the many major differences between the two animals indicate that they are not closely related. For example, the dolphin breathes with lungs, while the shark does so with gills. The dolphin has a mammal's four-chambered heart, while the shark has a two-chambered heart. The dolphin has a backbone; the shark, a skeleton made of cartilage, which indicates its relationship to cartilaginous fishes. The dolphin, furthermore, has mammary glands and some hair—unmistakable attributes of a mammal.

As these examples indicate, the fascinating and significant fact is that convergence occurs across the gulf of untold millions of years, across the barriers of species differences, as well as across thousands of miles. There is good reason why this happens: the environmental and physical forces mold the animals into similar shapes, especially those animals that are engaged in similar ways of life.

This suggests that some creatures on other planets will be found to be or to have been strikingly similar to our own. They are bound to have readily recognizable bodies, heads, and appendages like legs and arms. (See Chapter Five, "Dictates of the Environment: The Look of Life.")

But this does not imply that the whole universe is populated by man-like beings or animals that are mirror images of Earthly creatures. Convergence does not always lead to similar appearance. To occupy a similar niche, a creature on one planet would not have to exactly resemble a counterpart beast on another. The kangaroo, for instance, can be looked upon as the Australian counterpart of the deer and other hoofed grazers such as cows and horses. But aside from the fact that all those animals eat grass, there isn't all that much similarity among them.

A deer, such as the placental impala of Africa, runs away from predators on long and slender legs, while the kangaroo bounds away on its enormous hind legs. The impala and the cow defend themselves with horns, the kangaroo by rearing up on its tail and thrusting forward the sharp claws of its hind feet.

The kangaroo thus hints at how creatures on other worlds occupying similar ecological arenas may differ from our fellow Earth dwellers—even if they evolved from a somewhat similar sequence of ancestors.

A fascinating lesson can be learned from these Earthly examples of what to expect on other worlds. There is no reason to believe, for instance, that life elsewhere, given the opportunity, will not try to fill ecological arenas similar to those on Earth, producing animals both resembling our life and differing from it.

We can't expect exact duplicates of man and other Earthly beings, however, because, as Dr. Simpson and other students of evolution have argued persuasively, evolution is nonrepeatable and irreversible.

In his admirable book *Life of the Past* (Yale University Press, 1976), to be sure, Simpson urges the application of "a little common sense" to the dictum that evolution is irreversible. He notes, for instance, that it is *not* common sense to conclude, as some enthusiasts once did, that a large animal cannot be the ancestor of a smaller one, that an enlarged tooth cannot later be reduced in size, or that an attached animal cannot become free-living. All these things can happen and have happened.

Evolution does not repeat itself, in the sense that organisms never return to an ancestral condition. Horses that had become *smaller* in the course of evolution (and later grew large again) were different from their small remote ancestors. Whales and dolphins returned to the sea, but they didn't become fishes, despite their fish-like shape.

The irreversibility of evolution also means that if life were starting over again on a primitive Earth, neither man nor other animals would again evolve into their exact present forms. Too many unpredictable happenings contributed to the rise of man and other mammals now living. If, as Robert Jastrow has commented, to bear life a planet must squeeze through "a narrow gate" (in terms of being exactly the right distance from just the right type of star), then man just barely squeezed through a crack in the evolutionary door.

Some Paleozoic fish fortuitously developed lungs—not to become land dwellers but to cope with their own environmental problems, namely gulping air as they got around shallow ponds. Climatic changes, solar and cosmic radiation, and many other factors all influenced the mutations that led to higher animals. There were altogether too many random events in this history for it to be duplicated exactly on another planet. (See Chapter Five, "Dictates of the Environment: The Look of Life.")

"Would, for instance, man exist now if some Paleozoic fish happened to turn north instead of south?" Dr. Simpson has asked. "Or if that particular species of fish had not existed at all? Or had existed a million years earlier or later than it did?"

Answers to those questions are central to the larger query whether evolutionary processes are truly cosmic or merely terrestrial. At least they are central where evolution of mammals in general and man-like intelligent beings in particular is concerned.

Yet there is a lot more to this fascinating subject. Once life arises, given the impetus of proper energy supply, it can't help but evolve into more advanced intelligent forms, although not necessarily as intelligent as man. (See Chapter Eight, "Future Man.") And while evolution may be unrepeatable in the sense that exactly the same creature cannot be created twice, as we have seen, physical and environmental conditions can mold surprisingly similar animals evolving in a convergent manner in different places at different times.

If three types of wolves originated on isolated continents on Earth, all surprisingly close in shape and function, why not somewhat different man-like creatures evolving simultaneously in different settings on Earth? And why couldn't creatures on other planets evolve into man-like beings, given a similar sequence of events?

It probably came close to happening on those isolated miniplanets on Earth.

It appears, for instance, that the prosimians of South America could have been one such candidate. Not unlike small lemurs, and already living in the trees, the prosimians diversified and grew to become the monkeys of South America, equipped with prehensile tails. Given enough time and continued isolation, they might have stepped onto the path to becoming more or less human. Man's arrival on the scene has almost certainly eliminated that chance for good.

Perhaps the South American continent wasn't big enough for another man-like creature to emerge. Australia, for instance, being even smaller, yielded no possible recruits. The koala bear today hardly seems a possible contender, although nothing is really impossible, given the millions and hundreds of millions of years that evolution takes to build more complex beings from simpler kinds.

It could be that what was needed for evolution of man-like apes was the greater range and environmental variety of the Asian-African arena. It probably offered greater scope for adaptation and on a greater scale. We know that, about 20 million years ago, apes of many kinds had spread among the forests of Asia and Africa.

Gradually, from these apes, developed many subhuman species. They roamed the continent for a long time, before one kind, our ancestors, got the edge, about a million years ago. "We are the survivors," says Dr. Berrill. "In every group, but a few kinds survive from among the many."

But it's only within the past million years that those ancestral apes became more human-like, and only within the past one hundred thousand years that man attained his present form.

Man wasn't the only candidate on Earth, however, to become a highly intelligent creature. There are obviously alternative roads to high intelligence. Let's say that those Paleozoic fishes had not been forced to become land dwellers, that those shallow ponds never came into being, and that fish had remained in the rivers and oceans, without venturing onto dry land.

Legions of other candidates stood ready to emerge from the sea: insects, crustaceans, even the octopus. Any of them could have become much more intelligent creatures.

What leads to the kind of intelligence we, apes, and dolphins and whales have is an extreme form of mobility. Evolution of a good brain, whether in mammals, insects, or mollusks, seems to be tied up with a highly active, exploratory way of life. In that kind of life, a multiplicity of sensory impressions impinge on the brain simultaneously, developing an aura of excitement and, in turn, stimulating the development of small nerve centers into large brains and minds.

How creatures such as insects and mollusks could have evolved into intelligent beings will be explored in detail in Chapter Six, "Ocean Kingdoms and Insect Worlds." Here we want to consider what could have happened on Earth—*after* those Paleozoic fishes had given rise to amphibians and subsequently to reptiles and mammals. We assume that the fishes would have found their way onto the land, either by chance or through necessity.

We associate human intelligence and man's two-footed stance with his descent from the trees. It all began, supposedly, with the arrival of a shrew-like mammal at the base of the tree. This mammal began to climb trees, grew bigger, developed skillful hands, stereoscopic and color vision, and the beginnings of a big brain.

The wide-open savanna beckoned with its richer supplies of food, so the ancestor of man got off all fours, in the process freeing himself from the slavery of the sense of smell that chains so many animals to the ground.

Hunting and gathering food in groups, the ancestors of man set off on their million-year-long trek toward skyscraper cities and the atom bomb. Now the stars beckon.

Man may think that he was chosen for this role, but a look at the part of the mammalian evolutionary tree from which man descended shows just as easily another candidate for a highly intelligent being branching off: the bat.

Although both man and bat stem from the same shrew-like mammal, the bat developed web-like wings to fly instead of skilled arms and hands to jump from branch to branch.

The bat could have been pushed to the fore if geological and climatic changes on Earth favored a flying creature instead of a walking ape. If the savanna had never opened up, for instance, the bat might have been favored by natural selection over the ape, as shown in our illustration on the opposite page.

There is no reason why the bat could not have grown substantially larger than it is today and why it would not have developed a superior brain that would have led it to live in colonies, build simple dwellings, and perhaps even invent musical instruments.

In terrestrial bats, forelimbs and fingers have been restructured into strut-like supports for the thin membranes that form the wings. Only a short, clawed thumb remains for grasping, while the hind toes are used for holding on in the upside-down, resting position. A membrane also stretches between the two hind limbs; female bats use it to carry their young.

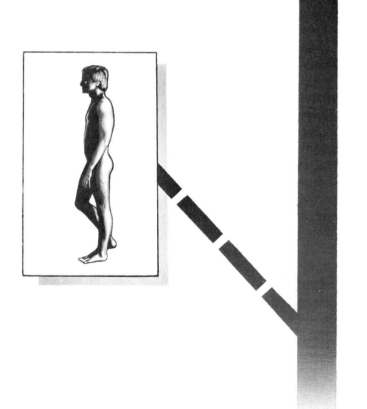

But even with their limited manipulative ability, some small Earthly bats build highly unusual shelters; some tropical bats, for instance, fashion tent-like shelters from palm leaves.

A bat-man like the one shown in the color insert, for instance, could do a lot of things man can do, and it could fly besides. It need not build a high-technology society—in fact, it probably couldn't, due to the limited ability to manipulate objects with its two-fingered hands.

Of all the primates, only man can manipulate the digits of the hand (the fingers) separately—apes and monkeys can't do it. This may be different on another planet, of course, where creatures could have started with more than five fingers. In that case, a bat would have enough fingers left over for a manipulative hand as well as a bat wing, to engage in more complex activities. The bat-man could speak through the nose, which would have evolved a broadcasting structure. It could weave baskets to carry its young and its food. As dolphins show on Earth, intelligent creatures can enjoy life without building technological societies.

Incidentally, bats haven't done too badly on Earth: of some forty-three hundred species of mammals, bats account for about 900, grouped in seventeen families.

We can't exclude even birds as a planet's dominant creature. In fact, birds on Earth seem to have had the stage to themselves following the extinction of the giant reptiles and preceding the first surge of the mammalian evolution. As a result, many of the original seed- and fruit-eating birds became very large grounded creatures, flightless because of their great weight, and displaying huge beaks, possibly for feeding on small reptiles, fish, and large insects. Without competing and preying mammals bothering them, birds could have become dominant creatures on Earth.

The birds could have become even more diversified than they are, and apart from other birds as predators, they would have been free to lay eggs in nests on the ground everywhere. In some birds, eggs would have been exchanged for placental wombs.

Insects could be around on a planet of the birds, perhaps in smaller numbers and fewer varieties, since they would be relentlessly hunted by the birds. It's possible to visualize a situation in which, early in the ascent of the birds, insects might have challenged them for the supremacy of the planet, only to be defeated and relegated to a furtive existence by their larger rivals.

If birds became dominant, their size could soar. Even without reaching that exalted position, some birds grew to impressive size. In 1980, scientists in Argentina found fossils of the biggest known flying bird so far. This was the teraton ("wonder" in Greek). It stood six feet tall, weighed 165 pounds, and had a wingspan of twenty-five feet. The teraton is believed to have died out only ten thousand years ago. By comparison, the largest flying bird now living, the Andean condor, weighs about thirty-five pounds and has a ten-foot wingspan.

Among impressive extinct flightless birds is the elephant bird, of Madagascar, which looked like a giant goose, standing twelve feet tall, and whose females laid two-gallon eggs. The moa, of New Zealand, also towered to twelve feet but looked more like an ostrich.

As the parrots demonstrate, birds have a capacity for speech. If birds developed brains capable of making use of speech communication, a high level of intelligence would probably result. (See color insert.) Even though birds would not be able to become skilled toolmakers and technologists unless they transformed their wings into arm-like appendages, birds on Earth display remarkable skills in building even by using their beaks and feet.

The weaverbirds are an example: They take blades of grass and parts of palm fronds, twist them around twigs, and weave them together with additional thin pieces into pear-shaped hanging nets with an elongated opening extending downward. The weaverbird works as skillfully as any basketweaver, holding the fabric with its foot and pushing loose ends into the meshes and tying them in with its beak.

Some birds have a highly developed sense of beauty. The male bowerbirds of Australia and New Guinea, for instance, build structures on the floor of the forest into which they try to entice females to breed with them. A nest-like structure with a roof of twigs and grass, the bower is beautifully decorated by the male with berries of various colors, blue iridescent beetles, yellow flowers, broken shells, and even glass beads and strands of tinsel if the bird can find them. The satin bowerbird of Australia even paints the interior of its bower with the juice of blueberries, which it crushes with its beak. The bird then picks up a piece of bark and, using it as a brush, spreads the juice on the walls.

Such activities hint at the possibilities inherent in birds. In spite of physical encumbrances, if birds grew bigger and smarter, they might be wondering about the stars.

Another possibility that could have shut off the development of man would have been the emergence of an intelligent reptile. What if the Age of Reptiles, which began 200 million years ago, had never ended? Why it ended, we really don't know, although there is no shortage of speculation. (See Chapter Five, "Dictates of the Environment: The Look of Life.") Had reptiles continued their development, mammals might have never been given a chance to evolve much beyond those tiny shrews.

Reptiles, on the other hand, could have become much more intelligent than they did. Even the fairly primitive reptiles that lived on Earth had the makings of more intelligent creatures. Some already walked on two hind legs, showing that it is possible to get to bipedal structure without first being some kind of monkey. And their forelimbs were at least as skillful as those of the squirrel. One Canadian scientist, in fact, has suggested that some of those early reptiles had enough skill to hurl rocks at their enemies. (See color insert.)

The still largely incomplete fossil record on Earth provides some tantalizing glimpses into how reptiles might have evolved into intelligent, upright-walking creatures. Writes Dr. Dale A. Russell, director of the paleobiology division at Canada's National Museum of Natural Sciences, in Ottawa: "One cloudy August day a Western Canadian rancher led me to the fragmented bones of a small dinosaur that died in Alberta 75 million years ago. The creature evidently walked on its hind legs, and possessed three-fingered hands in which the outer finger closed against the two others like a thumb. Its eyes were large and directed toward the front of the skull, suggesting a stereoscopic field of vision. Most significantly, the cerebral hemispheres of the animal were enlarged, exceeding those of any living reptile in relative size and equalling those of some living mammals. This small dinosaur was a manifestation of the widespread tendency of animate organisms to become more intelligent through geologic time."

Interestingly, in a recent Soviet computer simulation of evolution of land animals, the dominant creature that emerged was a highly active and intelligent tail-less reptile walking on hind legs and possessing skillful hands. Its descendants, the simulation further showed, were medium-sized upright walking reptiles that had fur instead of scales. As a matter of fact, a man-sized, fur-coated, mammal-like reptile, Cynognathus, apparently on its way toward something better and brainier, did live on Earth, but unlike in the computer simulation, walked on all fours. Our illustration on the opposite page shows an intelligent reptile taking man's place on the evolutionary tree.

Being cold-blooded, reptiles would have to be restructured to develop bigger brains. They would have to shed their tough skins for

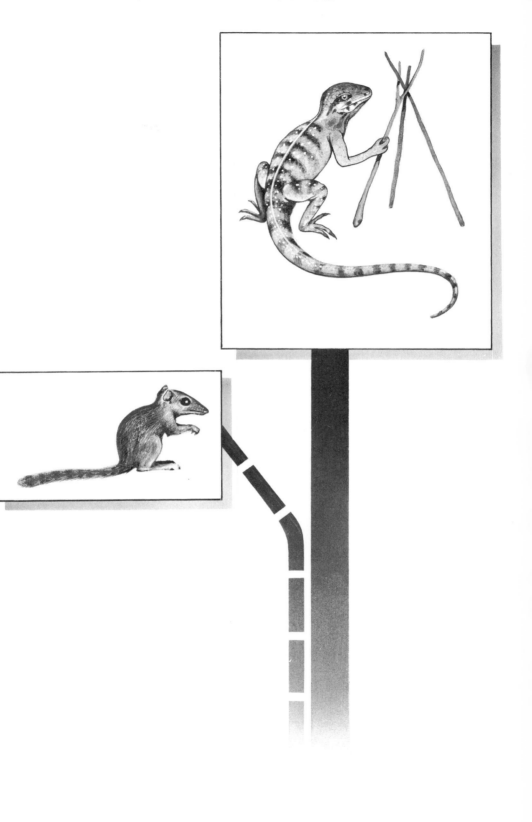

better temperature control. Unlike mammals, reptiles have no internal temperature regulators. They crawl or walk from shade to sun to control their body heat. Their scaly skins are not equipped to let the reptiles cool themselves by respiration. The scales are, in fact, designed for just the opposite purpose: to prevent the loss of water.

Cold-bloodedness may be an advantage in terms of the ability to subsist on relatively little food. Some snakes, for instance, coil themselves around tree branches and stay there for weeks at a time. But cold-bloodedness certainly is a disadvantage in terms of development of a large, intelligent brain, which demands a high rate of metabolism. To become highly intelligent, in other words, reptiles would have to have a much higher rate of metabolism. Put another way, reptiles would have to become more humanoid, or at least more mammalian in appearance. (See color insert.) From such a reconstructed reptile, a pseudo-human creature could evolve without the benefit of having climbed up and down the trees.

There is no reason why such a reconstruction of a reptile could not take place on other planets. Evolution is a great restructurer. It takes equipment and parts at hand, and by using unimaginable millennia, it keeps remolding existing parts and structures, adding on to them and using old parts in new ways.

The human ear is an outstanding example; it consists of parts that originally had little or nothing to do with hearing. The component parts of the patchwork of membranes, bones, and ducts that now constitute the ear date back to a balancing device in jawless fishes, to fish gills, and to skeletal elements in the throats and jaws of the first reptiles.

Underscoring once again the fact that nature takes advantage of evolutionary opportunities and works with the material on hand, as the end result we have the mammalian ear—a complex structure, but one that works.

Not only organs but whole animals have been so remolded into new shapes to exploit new environments.

The dolphins and the whales are prime examples. Both are mammals and both originally were dog-like landlubbers. They are believed to have descended from the creodonts, the extinct primitive carnivore suborder.

At a time when these primitive mammals began to branch out into bats, moles, and primates, dolphins and whales chose to become reentrants, mammals that went back to the sea.

Could they return to the land to become even more intelligent creatures than they are?

Putting evolution into reverse, it would take mind-stretching millennia, perhaps as long as 100 million years, to reshape the dolphin's fins into the legs they once were.

The embryos of dolphins and whales display four limb buds. Usually, the two hindmost ones are resorbed during growth in the womb. Occasionally, however, a dolphin or a whale is born with residual limb buds, showing that ancestors of the dolphins and whales had four legs.

It might be easier for the seal to return to full-time life on land to be reshaped into the dog-like creature it once was and that it more than faintly resembles today. The seal is not as intelligent as the dolphin; dolphins have been known to tease seals without leaving a doubt as to who is on top.

The dolphin's flippers might be too far gone to be turned into legs again. But if the dolphin were to return to land, its large brain might turn it into a planet's dominant creature.

"When you think that a fish fin became a walking leg and has now reverted to being a dolphin flipper, you can't say that a dolphin flipper couldn't in the course of a long time become reconverted into a walking leg, if the animal had a chance to get out of the water," says Dr. Berrill.

The dolphins, as shown in our drawing on the next page, would be forced to change their physical appearance to some extent if they returned, the same way reptiles became changed from fish, gradually changing their facial features to fit their new diet.

Some scientists believe that even with the dolphin where it is, it already represents a second intelligent species on Earth, perhaps even more intelligent than man. The idea, to be sure, has not been proved to everyone's satisfaction yet, but anecdotes abound about the dolphin's intelligence and human-like behavior.

There is, for instance, the remarkable story of Donald, the dolphin who hung around the Isle of Man, in England, in the early 1970s. Donald not only displayed the playfulness normally associated with his kind but also some mischief. Once, he started pulling a yacht out of a harbor by picking up its anchor. On another occasion, Donald got terribly agitated when a male scientist had placed a big rock on the lap of his female associate to photograph her on the bottom of the sea without the woman's floating up. Donald apparently thought

the woman was trapped and kept pushing the rock off her lap with his snout. Only after the woman scientist herself removed the rock, swam alongside her male colleague to a boat, returned to the bottom and put the rock on her knees herself, did Donald quiet down.

Donald displayed other intriguing traits. At times, he would swim to remote parts of the shore, where he lingered as if to listen to the rhythmic whispers of the little rocks being rolled by the waves.

Similarly, a variety of Pacific dolphins known as spinners are often seen performing what looks like a complicated aquatic dance in which two animals will jump into the air and spin in unison with fins touching, apparently doing the precision dance strictly for fun.

When Donald played among swimmers in shallow water, he was careful not to touch or push them, as if realizing that his three-hundred-pound bulk could cause damage.

The scientists observing Donald concluded that his quick and subtle reactions could only be compared to thought processes that are traditionally believed to be man's exclusive province.

In more systematic research aimed at creating an artificial language that may allow dolphins and humans to talk to each other, scientists taught dolphins to carry out specific tasks in response to as many as twenty-five "words" arranged in varying sentences. One of the dolphins, in experiments conducted at the University of Hawaii, learned to audibly name objects such as balls and hoops in the artificial language. Significantly, the researchers have concluded that dolphins can understand the abstract meanings of words; the animals respond correctly when the words are inserted in new sentences.

On command, a dolphin will carry a ball to the gate at the other end of the pool. The animals can discriminate between right and left, and on a command made up on the spur of the moment, they'll execute the proper new moves.

When one of the dolphins in Hawaii was instructed to push a person, instead of a ball, through a hoop, for instance, the animal unhesitatingly did so, suggesting to researchers an understanding of the generalized meaning of "through."

It is the dolphin's exceedingly large brain, on the average larger than man's, that fascinates many scientists. They suspect that it has other uses aside from mere echo ranging, which dolphins use to locate fish for food and to guard against predators such as sharks. Even philosophical attributes have been ascribed by some scientists to the brain of the dolphin.

"The fact that the porpoise,* without having enjoyed living in the treetops, without having become a two-legged hunter on the plains, has somehow managed to produce a brain whose complexity is comparable to that of man is a puzzlement," says Dr. Berrill. "It does show that we must not base everything on human experience. And, therefore, large, intelligent brains are possible without having gone through our story.

"I find it difficult to see why the porpoise needs such a large brain considering the restricted nature of its surroundings. We are stimulated by seeing and hearing everything from a very complicated universe. In the water, there is not much to hear except the noises you and your own kind can make. There is very little to see because one can't see very far no matter how good your eyes are. Yet they came up with this big brain.

"The question," adds Dr. Berrill, "is why did the porpoise get such a big brain and what does it do with it? In our own case, we are dealing with properties of the brain and mind which were never selected for by evolutionary processes, like music and astrophysics. These are fallouts. I don't think these are trivial happenings. I think this is what happens when evolution of this stuff we call a brain gets above a certain point. The universe represented by us and our kind is now becoming self-conscious and conscious of the rest of it."

Interestingly, the brains of dolphins and other cetaceans, such as whales, are enlarged in areas that deal not with the creatures' sonar but, rather, with memory, planned activity, and future contingencies. Comparison of man's and dolphin's brain shows that the dolphin has more cortex left over for the higher mental processes than we do. In short, dolphins have human-like brains, only bigger.

What's more, whales and dolphins, with their big brains, have been around a lot longer than man, perhaps for as long as 50 million years, while man's present brain dates back only 150,000 years. So they have had a lot more time to improve their brains.

In further support of the view that the dolphin uses its brain for purposes other than echolocation, the bat, which is equipped with a sonar system no worse than that of dolphins and whales, operates it with a relatively tiny brain.

One reason why dolphins and whales have large brains may be because sound transmission serves them in many diverse and fascinating

---

* The terms "dolphin" and "porpoise" are often used interchangeably, although the dolphin is larger than the porpoise. Both belong to the family Delphinidae.

ways, not only as a means of identifying objects and each other but also of expressing feelings in sound instead of body movements.

Where dolphin language—"Dolphinese"—is concerned, some scientists postulate that it is based on construction and processing of "acoustic pictures"—a structure entirely different from that of human languages, which are based primarily on visual and mental images.

So highly developed is the dolphins' sound-sensing that they can literally "see" through objects, using sound the way radiologists employ X rays to see through the body.

Sound waves carried in water penetrate skin, muscle, and fat and are reflected off air cavities and bones. Bouncing sound off another dolphin, a sending dolphin receives a diffuse picture of his friend's body contours. Superimposed more clearly on the contour are the breathing passages, including the lungs and air cavities in the skull, as well as teeth and bones. One dolphin can even see if another is suffering from air pockets in the stomach—suggesting how various senses might be developed in creatures on other planets in ways unattainable to us.

As for communicating emotions, the dolphins may be doing so by means of encoded sound transmission rather than relying on changes in facial expressions, vocal overtones, and other techniques employed by man. The dolphin's laughter—or his tears—may be represented by those encoded signals, a system of communication as alien as we can imagine and one that we will have to crack if we are going to have any success in deciphering the language of life-forms from other planets.

On the other hand, there are scientists who view the dolphins as not being much more intelligent than dogs. They suggest that such anthropomorphic activities of dolphins as rescuing drowning humans represent nothing more than an expression of the maternal instinct in female dolphins, which push their newborn calves up above the surface for their first breath of air. Furthermore, because they can't breathe underwater, dolphins will support an unconscious fellow above water until respiration starts again.

If dolphins are so smart, ask these scientists, why do they keep returning to the same Japanese waters year after year to be slaughtered by fishermen?

The dolphin dilemma may not be resolved until their complex language of clicks, whistles, and other noises is deciphered.

Attempts to talk with primates such as chimpanzees and gorillas via sign language haven't led to any significant dialogue yet; in fact, some scientists have lately begun to question whether those primates are really expressing linguistic abilities or merely their imitative powers.

Dolphins are far more intelligent than chimpanzees and gorillas; among other things, they learn tricks faster than the fastest monkeys and apes.

While we wait for that momentous event—the first real possibility of communicating with another intelligent species—the fact that such a species, with a brain that may be superior to ours, shares the planet with us is grounds for great hope in terms of evolution of intelligence in the Darwinian universe.

From all we know now about evolution of life on Earth and the evolution in interstellar space of molecules that make up life, to assume that we represent the only life in the universe would be to return to the egocentricity of the Dark Ages, which placed the Earth at the center of the cosmos and proclaimed man life's crowning achievement.

The variety of animal life on Earth, with more than a million species, which walk, crawl, hop, swim, fly, burrow, squirm, stay fixed to one spot, and range in size from malarial parasites one eighth thousandth of an inch long to whales more than one hundred feet in length, represents only a fraction of life-forms that have, or could have, lived on Earth.

The possibilities that have been exploited by evolution constitute only a small portion of those that could have been. About 250,000 species of fossil plants and animals have been named, described, and deposited in the museums of the world. This vast quantity of information, however, represents only a tiny fraction of life's diversity in the past.

"Known rates of evolutionary turnover make it possible to predict how many species *ought to be* in our fossil record," says David M. Raup, of the Field Museum, in Chicago. "That number is at least one hundred times the number we have found." That adds up to 25 million species—on Earth alone.

Multiplied by the billions of life-bearing planets circling other suns, the total number of cosmic species staggers the imagination. But as we will see in the next chapter, all those creatures are subject to universal laws that govern Darwin's universe, and as a result, we will easily recognize them as life.

CHAPTER FIVE

Dictates of the Environment: The Look of Life

How can we be so sure that creatures elsewhere will resemble our kind of life? Haven't we heard people argue that to think of life elsewhere as being similar to ours is to engage in a new kind of cosmic chauvinism?

It's not only nonscientists who automatically assume that life elsewhere will be unlike ours. Some scientists, too, have argued that extraterrestrial life won't be like any organism or machine with which we are familiar, because its evolutionary paths will have been laid down in environments different from those on Earth.

Both those laymen and those scientists are just plain wrong.

Look at what happened on Earth. On isolated continents thousands of miles apart, in environments that differed sharply from one another, evolution nevertheless created three wolves, two of them marsupial and the other placental, from ancestral animals that resembled the shrew and were only distantly related. And tens of millions of years apart, evolution poured the marine reptile ichthyosaur and the once-land-dwelling mammal dolphin into strikingly similar shapes to cope with their new environment of the sea. Examples of such convergence in shape, form, and ways of life are legion.

THE SHAPES OF LIFE

Where form, or shape, is concerned, the universal laws of physics and plain mechanics dictate it. To get around on land, a creature—especially an intelligent one—should stand fairly high above the ground to see where it is going and to see who may try to stop it from getting there. To fly, a creature has to assume the streamlined shape of a bird, or of a blimp in a more rarefied atmosphere. To swim, it has to be as streamlined as a fish, or nearly so. Throughout the cosmos, these dictates of the environment will be surprisingly similar.

Living beings everywhere will have to contend with gravity, friction, turbulence, drag. The greater a creature's speed the greater the shaping. Those physical constraints, for instance, mold fish, sea mammals, sea birds such as penguins, mollusks such as squid (which swim backward), submarines, airplanes, and rocket bodies (which must fly through the atmosphere). As Dr. Berrill puts it fittingly: "If you pulled a camel through the eye of a needle, it would come out as a thread—and so would anything else. I think this side of evolution is universal."

For those reasons, the creatures populating the Darwinian universe will be generally like us—yet different, too, just as a kangaroo is different from a deer or a cow, although the kangaroo also is a grass eater. It evolved in isolation in a different environment, as we have seen in the preceding chapter, and chose a different means of getting around and defending itself, indicating that alternative evolutionary paths exist for solving the same biological problems.

Still, we easily recognize the kangaroo as a member of the animal kingdom, both different from and similar to its counterparts on other continents, which were our own miniplanets on Earth.

THE CHEMICAL BASIS OF LIFE

Where internal chemistry is concerned, however, the similarities between us and our cosmic cousins start at a less visible and much more basic level: that of atoms and molecules.

"Many people feel that the moment you get outside the planet

Earth everything will be different," says Ralph E. Lapp, a noted nuclear physicist and author. "I know from personal experience that this isn't so. I once did an analysis of a type of radiation coming from stars and one of the most fascinating experiences is to take a picture of a star in terms of its line spectrum and to see that the wavelength of the hydrogen alpha line, for example, is precisely the same, whether the hydrogen is inside a star millions of light-years away, or whether it is being produced in a laboratory right here on Earth. The rules that apply here also apply a million light-years away—they are the same rules. The electromagnetic spectrum is the same on Andromeda as it is on Earth."

Just as the stars are made of the same elements that exist everywhere in the universe, so is the fabric of life woven of chemical strands that are the same everywhere and exist throughout the universe in about the same proportions. These life-giving elements—carbon, hydrogen, oxygen, nitrogen, phosphorus, and a number of others—furthermore, rank at the top of the list of cosmic abundance of elements. They make up the essence of all living things familiar to us—from a microbe to a blue whale—and they appear as constituents of stars and planets. Thus, literally we are children of the universe.

This scheme could hardly be an accident or a coincidence. It shows instead the universal unity of inanimate and animate matter, the fact that both are constructed of the same building blocks. There is thus no reason to expect beings elsewhere not to be made up of the same assemblies of atoms and molecules as is life on Earth.

What happens on Earth happens elsewhere. Inside the universal chemical laboratory, nature plays by definite rules. It puts together the same atoms and molecules everywhere. For this reason, no less an authority than Albert Einstein was convinced that there are no accidental occurrences in nature and that the law of causality is universal. One of Einstein's favorite sayings was, "The greatest miracle is that there are no miracles."

Science fiction writers and some scientists may speculate about fantastic beings on other worlds, but astrophysical and astrochemical observations clearly demonstrate that the underlying elements of life should be the same everywhere.

For the first time, we now have observational proof of this assertion, because we are fortunate to have acquired the ability to scrutinize the universe as if we were sitting inside a huge chemical laboratory—which the universe is.

The great radio telescopes, with their associated detectors, have been pointed at the skies and have revealed the presence in interstellar space of the same molecules that make up life on Earth. As Dr. Ponnamperuma has remarked poetically: "The whole universe is trying to make life." And it's our kind of life.

CARBON, THE HUB OF LIFE'S WHEELS

At the center of life on Earth—and elsewhere, it now appears certain—is the carbon molecule. The reason for this is that selection of the best available materials is inherent in the way nature works. Just as biological evolution chooses the best possible genetic combination for the survival of an organism—or does so in most cases—so does nature select the best basic materials as life's building blocks.

There is no chauvinism involved in arriving at this conclusion, only extrapolation on the basis of what we know and see.

Life as we know it is made up of four of the most abundant elements in the universe: hydrogen, oxygen, carbon, and nitrogen. Most abundant in the universe at large—but not on Earth.

On Earth, evolving nature chose those four basic elements, ignoring many others that were readily available.

There are rather striking contrasts in the abundance of elements in the universe as a whole and on Earth. Hydrogen and helium constitute nearly 100 percent of the universe. In the rock of the Earth's crust, on the other hand, 98 percent of the atoms are represented by oxygen (47 percent), silicon (28 percent), aluminum (8 percent), iron (5 percent), and smaller amounts of calcium, sodium, potassium, magnesium, titanium, hydrogen, phosphorus, manganese, fluorine, strontium, and sulfur. The presence of all these elements in both animate and inanimate matter shows the close interaction of environment and life.

In constructing living beings, though, it appears that nature was not guided by the local availability of particular elements but by some other yardstick. If we compare the availability of carbon and silicon, for instance, we find that, in the rocks on the Earth's surface, silicon outnumbers carbon by the surprising ratio of 135 to 1! Silicon is situated just below carbon in the periodic table of elements, and like carbon, it displays a tendency to bind four electrons, creating four so-called covalent bonds.

Why, then, is life on Earth based on the relatively rare element carbon and not on silicon, which is available in much larger amounts?

Science today is prepared to answer that question—at last. It turns out that the matter of which life is made demands the fulfillment of two conditions in the construction of its smallest building blocks: minimal size, and stability when one to four electrons are bound together.

Carbon, hydrogen, oxygen, and nitrogen passed the test. Silicon did not.

On the atomic level, carbon is the hub in life's wheels, because it has more "spokes" where other molecules needed for life can fit with appropriate strength and flexibility.

Carbon has four electrons in its outer shell, which can be joined by another four. This symmetry makes for the ease with which carbon atoms combine with each other and with other molecules. This is what places carbon in an exceptional position in the chemistry of life. Carbon unites easily with other atoms to form long polymer chains. Combined with the constituent atoms of the water molecule, the two hydrogen atoms and one of oxygen, for instance, carbon forms carbohydrates.

The versatility of carbon is unmatched. Carbon enters into more compounds than all the other 102 elements put together. Carbon can combine with hydrogen, for instance, to form such diverse compounds as asphalt, vaseline, sugar, and fat. Combined with hydrogen, oxygen, nitrogen, and sulfur, it forms proteins. Carbon comes as coal, the graphite used in pencils, and diamonds. In short, carbon's ability to combine with other carbon atoms and atoms of other elements knows few limits. Life needs long molecular chains.

No other atom approaches carbon in those abilities. Superficially, silicon looks like a candidate, and a generation of older science fiction writers have built up a whole population of silicon monsters, often with ammonia flowing through their veins instead of water. But the choice of silicon as one example of unearthly chemistry is highly unlikely if not impossible.

Silicon molecules really don't have the properties of organic molecules: they don't react or combine easily, and they lack double and triple bonds, the attributes that allow biochemical reactions to take place. Under ordinary conditions, polymers made of silicon are too stable to become ingredients of life. On Earth, for instance, silicon is bound up with oxygen in rocks and sand.

While silicon, like carbon, can combine with four hydrogen

atoms, it produces, of course, not methane but the gas silane. And it can do so only under extremely high temperatures. Under the best of circumstances, the bonds between silicon atoms are only half as strong as those between carbon atoms. Therefore, long molecular chains would be difficult to form from silicon.

Another attribute of silicon that reduces its candidacy is its affinity for oxygen. It forms silicon dioxide—quartz—at normal (Earth) temperatures; quartz is not exactly suitable as a component of life. It is a solid unless heated to the extremely high temperature of 1,372 degrees Fahrenheit (744 degrees C), is highly insoluble, and is difficult to decompose. Silane, the silicon-hydrogen combination, on the other hand, catches fire when exposed to an oxygen atmosphere.

So while carbon combined with oxygen results in carbon dioxide, a molecule that interacts easily with other components of life, silicon, in contrast, removes itself from such interactions. Silicon would hardly seem to be a promising backbone of life.

"After four billion years of evolution," says Dr. Ponnamperuma, "we have silicon only in structural molecules, never in functional. Carbon is used everywhere. If evolution could have used silicon, it would have made something out of silicon—there's a lot of silicon around."

All in all, silicon appears to be more suitable to be made into rocks that are walked on instead of into organisms that walk on them. Silicon is far too stable to be a functional component of life, which is an unstable state of matter.

As a result, silicon monsters are no longer taken seriously by most people. Can anyone contemplate, with a straight face, a silicon creature exhaling cubes of quartz? The next question, then, is, is there any reason to believe that life might be based on nitrogen, boron, or any other atom?

"I think it's very unlikely anywhere," says Dr. Norman Horowitz, professor of biology at the California Institute of Technology. "It's unlikely if you assume, as I think most people do, that in order to have a living system you have to have a system that is capable of storing and replicating large amounts of information.

"All our genetic heritage is just information that the species has acquired over hundreds of millions of years of evolution. The results of this experience are encoded in our genes that tell us how to live and without this we couldn't survive. The essence of the living state is to have access to these large quantities of information, to be able to replicate this information, and be able to transmit it to the next generation. And if this is based on molecular complexity, then you have to base your life on an atom that is capable of giving rise to very large and very complicated yet very stable molecules. If you look at the atoms available, carbon is obviously the choice for this role."

WATER, CARBON'S COLLABORATOR

Life needs a solvent, too. Water is carbon's indispensable companion where life is concerned. "Water," says Itchiaque Rasool, a NASA scientist, "gets everything going." Consisting of two molecules of hydrogen and one of oxygen—both gases—water is nevertheless a liquid.

By remaining liquid over a large range of temperatures, water permits chemical reactions to take place. Water dissolves other chemical compounds, allowing organisms to use it to carry nutrients in and to eliminate the wastes.

In the absence of water, chemical reactions do not take place, or occur so slowly that their progress cannot be detected. Water is the universal catalyst. It speeds up chemical reactions without itself being changed.

Without large bodies of water, carbon-based life could not have evolved. "Anywhere there is liquid water," says Dr. Ponnamperuma, "life will probably arise."

On the basis of cosmic abundances, carbon dioxide can be predicted to be a major constituent of early atmospheres of Earth-like planets. And it is water that converts atmospheric CO_2 into solid carbonates. The absence of water on Mars and Venus, as we shall see shortly, kept the CO_2 in those planets' atmospheres.

Water is the most abundant constituent of life. Any living organism is in reality a mass of water encased in a protective bag: inside it, chemical reactions proceed to maintain life and to assure its perpetuation. In humans, water makes up 70 percent of the body's weight. In all organisms, water helps to regulate body temperature.

In its vital ability to handle other chemicals, water is far superior to both ammonia and methyl alcohol, the two other solvents that have some of water's unusual properties.

Unlike ammonia and methyl alcohol, bodies of water freeze only on top in winter, allowing life in ponds, lakes, and seas to continue even if the surface is frozen. Oceans of liquid ammonia, hydrogen cyanide, hydrogen fluoride, hydrocarbons, or molten salts, on the other hand, would all freeze from the bottom up. No life could continue in bodies of such frozen solids—at least no higher life.

Most significant of all, water provides a built-in shield against ultraviolet light from the sun. Even as the bonds of some water molecules are broken by ultraviolet radiation, some of the oxygen atoms so released will rise into the atmosphere to link up to form ozone, which protects life on Earth.

Ammonia could offer no such protection against ultraviolet radiation, because it would produce nitrogen instead of oxygen atoms.

On Earth, the ancient reducing, or hydrogen-rich, atmosphere was also transparent to deadly ultraviolet rays. To protect themselves, the first organisms lived in watery depths. Later, photosynthetic bacteria evolved and began to release oxygen into the atmosphere, allowing the protective ozone layer to be formed. (See Chapter Two, "The Curtain Rises.")

Some science fiction writers and a few scientists still hold out for ammonia as a substitute for water, with silicon or boron taking carbon's place. Those combinations, though, really don't work out. Any boron-ammonia chemistry, for instance, if it worked, would provide such a limited biological function that scientists even have trouble visualizing it.

Liquid ammonia, furthermore, is far more sensitive to temperature extremes than is water. It can exist in a liquid state only within the range of −28 degrees Fahrenheit (−33 degrees C), its boiling point, to −98 degrees Fahrenheit (−72 degrees C), its freezing point. This is less than half the temperature range of water. A convenient place for a planet with a temperature range to suit ammonia in our solar system would be between Mars and Jupiter—not a very inviting neighborhood for life, especially intelligent life.

What's more, for a certain amount of ammonia to be present, more water must be around than ammonia. The tables of cosmic abundance indicate that. There is far less nitrogen around than oxygen, the respective constituents, with hydrogen, of ammonia and water.

Interestingly, water has served not only as the all-important lubricant and constituent of life, but it also was instrumental in creating the land we walk on. Water reacts with many other hydrides—binary compounds of hydrogen such as those of silicon, aluminum, and magnesium. Such reactions yield the products constituting most of the Earth's crust.

On the other hand, some of the so-called covalent hydrides—methane, ammonia, hydrogen sulfide, and phosphine—are stable in water. They underwent a breakup by ultraviolet light on the primitive Earth, yielding complex compounds such as hydrocarbons, carbohydrates, and phosphates—all biological compounds.

Thus water is a basic requirement of life not only as a solvent for biochemical reactions but also because it brings about a separation and synthesis of the components of both geology and biology. There is no reason to believe that water does not play this role on other planets too, for the abundance of water occurs thanks to the most abundant element in the universe: hydrogen.

The abundance of hydrogen, in turn, bodes well for a superabundance of life in the universe. For wherever oxygen is present in addition to hydrogen and the temperature is not too high, water will form. Says Carl Sagan: "It's just in the chemical cards."

It's hard to escape the conclusion, on the basis of these facts, that the overwhelming preponderance of life throughout the universe will be based on carbon-water chemistry. The observable facts dictate this conclusion: the abundances of the various elements appear about the same everywhere astronomers turn their optical and radio telescopes. Silicon chemistry makes up silicates—rocks and sand. Carbon, in contrast, appears not only as the principal constituent of living things but is also found in meteorites, comets, in intersteller space—where no silicon is detectable. The exceptional properties of carbon and of the other elements that constitute life on Earth make them the prime candidates as chemical constituents of life everywhere in the universe.

"With all the knowledge we have today," said Dr. Ponnamperuma in a recent interview, "we are willing to say that eventually we will define life as the property of the carbon atom.

"I teach an undergraduate course in chemistry for nonscientists. And the first thing I tell them is that a professor in the constellation of Andromeda will be teaching the same course."

A question about chauvinism brought this rebuttal: "There is no carbon chauvinism—the number of elements in the universe is the same; it's the same periodic table."

Dr. Ponnamperuma went on: "We can even go further than that

and talk about the DNA molecule. We find guanine and the other bases ideally suited to be nucleic acids. So it's not only the carbon atom but also the molecule. I'm willing to bet my bottom dollar that if we find life somewhere else, it's going to be a nucleic-acid, protein life."

Then he returned to "carbon chauvinism": "So when one talks about carbon chauvinism, where the interstellar hydrogen cyanide and formaldehyde are in the interstellar medium—we didn't put it there. Those are the very things we get in our laboratory mixtures imitating the primordial atmosphere. So the whole picture fits in very beautifully. At the chemical level, it's a very hopeful, optimistic situation. At one time, people thought that organic chemistry was something very unusual. To me today, it looks like God is an organic chemist.

"On the chemical level, we have a great deal of evidence to suggest that life elsewhere is probably very, very similar—I'd be willing to say almost identical to ours. On Earth, for example, the digestive processes of an elephant are no different from that of an *E. coli* bacterium—the cycles are the same. So on the chemical level, the components and the processes are similar."

THE PHYSICAL APPENDAGES

Not only the structures and the processes but also the appearance of creatures everywhere will be determined by the same natural laws. The reason is simple: only a limited number of biological engineering solutions are possible to solve problems that face living organisms. That's why unrelated kinds of animal life have nevertheless come up with the same inventions: legs, heads, wings, camera-type eyes, and red, oxygen-carrying hemoglobin, among others.

Heads are a necessity. "If you want to move on land or underwater in any direction," says Dr. Berrill, "you inevitably become stretched out. The sensory structure will lie in front. There is a natural tendency to develop a head so you can either sense what you are coming up against or maybe anticipate it."

And so it is with many other features and appendages of living beings—they are there because nature found them to be most suitable for the purpose intended.

These limitations may even give evolution its direction—if it has a

AN INTELLIGENT REPTILE   *An environmental change with which they couldn't cope apparently caused the extinction of dinosaurs and other large reptiles on Earth. If reptilian development had continued, however, highly intelligent reptiles (above) could well have evolved on Earth, and could evolve on other worlds. Upright-walking reptiles with good-sized brains and skilled, three-fingered hands, the dromaeosaurids, have lived on Earth.*

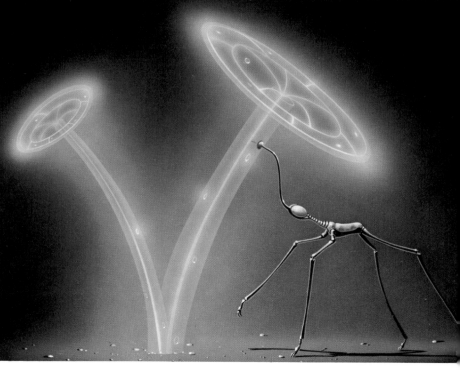

INTELLIGENT INSECTS AND RECHARGEABLE
ALIENS   *A variety of creatures, ranging from
the faintly familiar to some that at first glance
appear puzzling, may populate distant worlds.
Insectlike creatures can be expected to dominate
many planets; on Earth, insects and their marine
ancestors constituted the first wave of animals
that conquered the land. Insectlike creatures
elsewhere can be expected to engage in even
more complex activities (above) than do their
Earthly counterparts. On the left, an insectlike
creature withdraws chemical energy from a
fireflylike plant that processes energy chemically
instead of photosynthetically. At right, an antlike
creature has decorated itself with a berrylike fruit
hung around its neck to indicate its exalted rank.*

INTELLIGENT BIRDS AND BATS *Birds, as seen above, given the right opportunity and circumstances, could have taken the place of mammals, as well as continuing as birds. Bat-men, below, could have become dominant creatures on Earth. Both men and bats stem from the same ancestor, a shrewlike mammal.*

The reason man emerged was the opening of the savannas due to geological and climatic reasons. Had the grassy plains never appeared, the bat would have been favored over man to emerge as the Earth's dominant animal. That sequence could happen on other worlds.

THE WATER PLANET would boast strange, fishlike organisms with siphons for propulsion, and octopuslike creatures. In addition, counterparts of squid, clams, and oysters would be found, as would spiny echinoderms like sea urchins and starfish. Mollusks and echinoderms are likely to make up the indigenous population of any water planet.

MARSUPIAL MEN—*Humanlike creatures can develop elsewhere from koala-bear-like ancestors. Marsupial females are equipped with pouches in which they rear their prematurely born young—a big advantage for evolution of large brains. Marsupial manlike creatures like those in our illustration could easily have evolved on Earth if marsupials had dominated the other theaters of life as they did Australia.*

direction. At least, in the view of many scientists, such limitations in the number of possible paths that can lead to the solution of a biological need were probably responsible for progress from fish to reptile to mammal to man. Evolution, in short, molds the plaster of life to fit the cast prescribed by the environment.

On other planets, the same pressures of natural selection will be at work, although obviously with variations on the main theme. The kangaroo, the koala bear, the marsupial wolf, the giant moa of New Zealand, the tree sloth, and the armadillo all suggest the possibilities for variation.

The main theme, though, is unlikely to be violated. Creatures elsewhere, for instance, will get around in ways similar to ours. Bipedal leaping and jumping evolved repeatedly and independently on isolated continental masses. So did walking on two hind feet among differing animal species: reptiles, birds, primates. We are not going to encounter beings that get around on wheels instead of legs, because tissues can't stand much friction, and without an axle a wheel doesn't do much good. Unless, of course, the whole animal acts as a wheel, and there are one or two examples of that type of locomotion on Earth. A number of chipmunk-like creatures roll tail-over-head to get away from their enemies—especially when going downhill.

Similarly, the same physical laws will dictate some major similarities between life on other planets and that on Earth. For mammals, as well as for amphibians and reptiles, one head, two eyes, and four limbs are ideal—at least four limbs in our kind of gravity. Equally ideal is the bilateral-symmetrical form of body, comprising two halves.

In the sea, such creatures as the octopus and the starfish took on a radial form which enables them to reach for food coming in from any direction, or alternately pursue their prey in any direction, with an economy of body motion. Yet on land even an octopus would find walking on all fours more convenient than lumbering along on more legs. The pitter-patter of more than four feet may work for insects and arachnids, but larger organisms would stumble all over themselves. Once again, nature favors simplicity and centralization.

The tendency to reduce the number of operating units and increase their efficiency means that the chances are good that advanced life-forms elsewhere will walk on either two or four feet, although, as we shall see later, strong gravity may call for more legs if the complications that come with that can be solved.

Physical requirements will dictate a similarity in sensory organs in totally unrelated species. A striking example on Earth is the development of camera eyes in mammals and in cephalopods such as the octopus and the squid.

As our illustration shows, although different in detail, the eyes in both cases exploit the same principle, unlike insect eyes, which consist of rods. The eyes of the higher animals are called camera eyes, because, as in a camera, the lens focuses an image of the object being viewed onto a plane of light receptors, which take the place of film. The pupil of the eye of the octopus is a horizontal bar, rather than a circle, but that's not a very big difference although it does suggest some interesting variations within a general framework.

Nature discovered those principles long before man did, and applied them to two radically different and only very remotely related organisms, because it's the only way to construct a nearly perfect eye. Again, because all worlds are governed by the same physical laws, parallel developments elsewhere seem inevitable.

Two eyes, especially as good as camera eyes, are sufficient although the presence of a third eye, on top of the head, cannot be entirely excluded. In Earth-based vertebrates, the vestige of the third eye is preserved as the pineal gland, in the center of the brain. It was designed for letting the light in through a sort of a translucent window, not an image-forming eye but a light-sensitive eye that told the animal whether it was getting dark or light or whether a shadow—perhaps of a predator—was going over.

In addition to camera eyes, other forms of vision, particularly in lower creatures, are possible, of course. On Earth, biological light-sensing mechanisms start with the simple lens, in single-celled bacteria, which concentrates light on a group of light-absorbing molecules behind the lens. This merely informs the creature whether it's light or dark. The range extends to the more complex eyes of invertebrates, which form no image but can tell that motion is taking place in front of the eye.

With an increase in the number of light-sensitive cells, the eye can receive an image. In the camera eye, one lens does it. In the compound eye of insects, however, another principle is at work. Insect eyes consist of a large number of tubes that serve as independent light units. The sum total of the light received produces a rough picture of the object being seen.

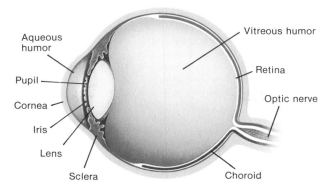

HUMAN EYE

Aqueous humor

Pupil

Cornea

Iris

Lens

Sclera

Vitreous humor

Retina

Optic nerve

Choroid

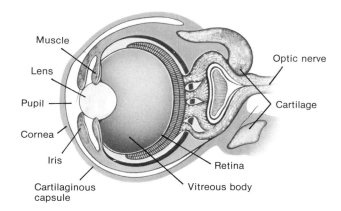

EYE OF A SQUID

Muscle

Lens

Pupil

Cornea

Iris

Cartilaginous capsule

Optic nerve

Cartilage

Retina

Vitreous body

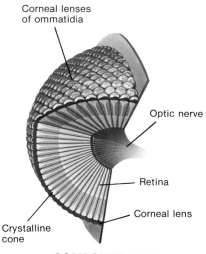

COMPOUND EYE (of insect)

Corneal lenses of ommatidia

Optic nerve

Retina

Corneal lens

Crystalline cone

Thus, recognizable camera eyes can be expected to be part of the equipment of creatures on other worlds.

The same would be true of such other sensory organs as ears, noses, and mouths, although perhaps not in the same proportions as in our animals.

THE BARRIERS OF SIZE

The sizes of extraterrestrial creatures will also resemble those of ours, because at the level of the organism there are constraints on size no matter where the animal lives. Not only gravity but skeletal and muscular systems will determine the optimal sizes of land organisms. In the oceans, such limitations are more relaxed, as seen in the example of the blue whale, at a hundred feet and a hundred fifty tons, the biggest creature on Earth. But there are limits in the ocean, too.

One limit is imposed by the organism's smallest unit, the cell. Whether it's a mouse or a whale, the individual cell is structured the same and is of the same size. So the bigger an organism the more cell "mouths" are there to feed—each has to be supplied with oxygen and amino acids, and each has to be cleansed of carbon dioxide and other wastes. The larger an organism the more complex becomes the distributing system.

Increasing a creature in size without limit would be equivalent to letting a city grow to an uncontrollable size—eventually, it's bound to strangle because of the clogged communications and transport lines. So even in the sea there are built-in limitations to the size of creatures. The biggest whales in our oceans have probably reached the limit in size; they are dependent on the tremendously abundant microscopic krill life in the waters off the Antarctic. The other limit probably is just how much blood a whale's heart can pump through the huge body. This is something that is not likely to be different on another planet, even on a larger one than the Earth, because the same physiological and physical restrictions will apply.

LIFE ON A DIFFERENT PLANET

The appearance of the creatures' external shapes, to be sure, will depend on the physical parameters of the planet where they live. A planet exactly duplicating the Earth in all its attributes—size, rate of rotation, precise distance from the sun, inclination of the planet's axis, and many others—probably would be as unlikely as would be the existence of an evolutionary double of man.

"It's more likely that you'll get intelligence in different body shapes," says Dr. Berrill. "You would not necessarily want to go up and embrace one of those creatures."

There may be a lot of Earth-like planets circling distant suns, but they won't be exact copies.

Changing just a few simple physical parameters of a planet would lead to profound differences in the appearance and types of life-forms there. A change in the spectrum of sunlight received, for instance, or a very rapid spin rate, would change the type—and color—of vegetation, aside from affecting animal life.

Plants elsewhere, incidentally, need not be green but can be of almost any color. Even on Earth there are purple bacteria. The reason why our plants are green is that they absorb a particular part of sunlight and reflect other parts.

Life on Earth has adjusted to a remarkable variety of conditions and extremes, and it will do the same elsewhere. No matter where we look on Earth, we find life—from the top of Mount Everest to the darkest ocean depths, several miles below the water surface.

A type of bacterium thrives in sulfurous hot springs by synthesizing heat-resistant structural materials. On the bottom of the ocean, strange, worm-like creatures crawl in eternal darkness at atmospheric pressures one thousand times that of normal pressure at sea level. To light up the ocean depths in their immediate vicinity, many dwellers of the deep produce their own light: bioluminescence.

To escape desiccation, many microorganisms enter into a spore stage, in which the water content of the cell is reduced and the cell wall or the spore coat becomes relatively impervious to water.

Microbial life has been detected at temperatures higher than 212 degrees Fahrenheit (100 degrees C), below −32 degrees Fahrenheit (−35 degrees C)—above the boiling point of water and far below it—and at pressures exceeding eleven hundred atmospheres. Laboratory experiments show that certain enzymatic reactions occur above 212 degrees Fahrenheit (100 degrees C) as well as at −45 degrees Fahrenheit (−43 degrees C).

Microorganism can even exist inside porous rocks in both hot and cold deserts. The organisms colonize a thin layer of rock a few millimeters below and parallel to the surface crust. In these protected niches, the microorganisms develop favorable conditions for life.

In hot deserts, only simple prokaryotic organisms, single-celled bacteria without a nucleus but with only a looped chromosome, live inside rocks. In icy Antarctica, on the other hand, eukaryotic algae and fungi, which are more complex organisms—with nucleus—and more closely resemble cells of higher animals, have chosen this way of life. They form a primitive, lichen-like association. In hot deserts, the main source of water appears to be dew, while in the Antarctic it's melting snow.

Furthermore, a multitude of microscopic life-forms have adapted to exploit a large part of the Earth's surface where liquid water is not available for extended periods of time. These organisms possess two properties that allow them to survive.

First, they have a physiological ability to switch their metabolism on and off. And second, they excrete metabolic products that modify the external microenvironment of the cell.

Some lichens, for example, modify the rock on which they live by boring into it. The microscopic tunnels created by the lichens' dissolution of minerals contributes to the porosity of the rock and thus to its water-holding capacity—in a remarkable demonstration of how even "unintelligent" life can change its environment.

So it would be a mistake to underestimate the ability of life to occupy niches that may seem preposterous to us. "I suspect," says David M. Raup, dean of science at the Field Museum of Natural History, in Chicago, "that the products of an independent evolutionary system might show some real surprises and might indicate that some of our 'optimal' solutions are not quite as optimal as we have thought."

Another planet's dense atmosphere, for instance, may appear just as inviting a habitat to the airborne Man-of-war-like creatures shown in our illustration (opposite) as our oceans are to fishes, mollusks, cephalopods, dolphins, and whales.

In such an atmosphere, incidentally, creatures would have smaller ears, because sound would be easier to detect. In a rarified atmosphere, in contrast, they would have larger ears.

The life-bearing planet would have to be of the right age to bear intelligent life—probably at least as old as the Earth, unless external conditions, such as nearness to its sun and consequent higher temperatures, have resulted in faster chemical reactions and more rapid evolution of life.

LIFE ON A SMALL PLANET

The size of a planet is another vital characteristic. Make a planet too small and it loses all or most of its atmosphere, as Mars did. Make it too big and it becomes a hydrogen-rich giant like Jupiter.

Life, to be sure, can exist on a planet smaller or larger than the Earth. Some scientists calculate the upper limit at three times the size of the Earth and the lower at two thirds the size.

With change in size, a planet's gravitational pull will change, of course, altering the shapes and features of life. On Earth, all objects are subject to a gravitational intensity measured as "g" ("g" being an abbreviation of gravity).

On a smaller planet, weaker gravity would release both trees and animals from gravitational restraint and make them shoot upward to become taller and more slender.

To visualize just what effect gravity could have on the forms of creatures on a planet with an extremely weak gravitational pull, it is instructive to look at our oceans. Life in the sea, of course, is not independent of gravity, yet gravity there is not a major determinant of shapes. The otherworldly forms of various fishes and other creatures in the sea show the possibilities: shapes of blowfish, flatfish, and the incredible lion fish, which looks like an utterly hopeless structure.

On land in such a setting, trees would be tall and spindly, and plants could assume spectacular and unusual shapes, growing huge, balloon-like appendages and extremely large leaves.

On a smaller planet, a giraffe-like animal could have a neck twice as long as its terrestrial counterpart, and trees could soar to heights of five hundred feet and more, as shown in our illustration (opposite).

LIFE ON A BIG PLANET

On a large planet, the opposite effect would be seen: animals getting squatter, their legs and necks getting thicker.

On a big planet with gravity five times Earth's, a one-hundred-sixty-pound person's weight would increase to eight hundred pounds. His ankles would be subjected to a tremendous pressure, so his bone structure would have to be changed to an elephantine one. He would need a much larger heart, too, since it would have to respond to the dynamics of pumping blood five to six feet above ground.

Shapes of creatures would be changed drastically. In experiments sponsored by NASA, young chickens were raised on a centrifuge that exposed them to three to four g's for their entire life-span. The birds responded by developing in unusual ways. Not only were their legs thicker, as would be expected, but in many birds the neck became bent backward, with the head resting on the animal's back.

It's also probable that higher gravity would have an effect on the atmospheric composition. This, of course, could happen without a change in gravity. If our imaginary planet's atmosphere contained only one third the amount of oxygen found in our atmosphere, for instance, that would call for much larger lungs—an attribute of the Indians living high in the Andes.

Such restructuring in its extreme form would result in a deformation of the present human and animal structures, so that creatures on such a planet would appear to have mammoth chests, tough bone structure, and somewhat squarish faces. Such relatively large changes would come from altering only two physical parameters on a planet. One possible set of inhabitants of such a world are shown in our illustration.

Some other interesting possibilities could develop on a large planet. To get around easier, creatures may become smaller, looking somewhat like their overweight counterparts on Earth. Tree-dwelling creatures would have to be small and fast, since there would be a premium on fast movement. In time, that could lead to the emergence of a race of "little people," possibly only two feet tall.

Another possibility that can't be completely excluded is for creatures on a big planet to have six legs instead of four, to distribute their weight better, and then perhaps to have two arms.

In Earthly fish, three pairs of fins could have formed as readily as two. "Two pairs of fins evolved for much the same aerodynamic reasons that airplanes have two pairs of wings and birds have a fan tail to function as the rear set," says Dr. Berrill. "On land two pairs of legs allow a greater stride than three pairs, and the best runners—men, ostriches, bipedal dinosaurs—have only one pair. Large creatures with more than four legs create headaches anatomically."

Still, while centaurs are out of the question—such an anatomy simply doesn't work—perhaps smaller creatures with more appendages are possible; we do have such creatures in the seas: crabs and lobsters.

THE COLOR OF THE ALIEN SKIES

The composition of other life-bearing planets' atmospheres may be different, of course. But to allow carbon-and-water-based life to evolve, it shouldn't be too different. On Earth, life began in a hydrogen-rich atmosphere but didn't start its explosive upward march until an oxygen-rich atmosphere was created for reasons explained in Chapter Two, "The Curtain Rises." As far as we can see, therefore, an oxygen-rich atmosphere is an absolute necessity for higher forms of life—whether they live in the water or on land.

Surprising as it may sound, or maybe not so surprising, considering the universal nature of physics and chemistry, life-bearing planets elsewhere will be similar to Earth even down to such details as the color of their skies. Skies on other planets will be just as blue as ours, even if their atmospheres are somewhat different in composition. The reason is that the color of the sky has nothing to do with the chemical composition of an atmosphere. It has to do, rather, with the size of particles in the atmosphere. The sky will be blue if the atoms in the atmosphere are small compared to the wavelength of light from

the local sun. But if the sun shines through larger dust—or smog—particles, the atmosphere—or a sunset—appears reddish.

On a smoggy day in Los Angeles, even the daytime sun often looks red, because large particles scatter the longer wavelength toward the observer. Similarly, the sky of Mars has a reddish cast because fairly large dust particles remain suspended in the planet's atmosphere.

When the particles in the atmosphere are small, on the other hand, shorter wavelengths are scattered more and also are scattered more homogeneously. The result: blue skies.

SOME SURPRISES IN STORE

Naturally, there are going to be surprises. Life has produced some startling creatures among the millions of species that have lived on Earth (see illustrations in Chapter Seven, "Life Beyond Darwin") so why shouldn't we expect surprises in the innumerable abodes of life in the universe?

The variations on the cosmic theme may include faster or slower animal metabolism, for instance. Slower metabolism on a planet farther from the sun than is the Earth conceivably could extend those creatures' life-span, but it may limit the development of a big brain.

We also know that chemical processes are speeded up by a simple increase in temperature. An increase of only 10 degrees F doubles the activity of most chemical reactions.

Conversely, on a planet closer to its sun or with a higher surface temperature obtained through a somewhat different atmospheric composition, body temperatures could be high, making the creatures into faster thinkers than we are—but with a price to pay: a drastically reduced life-span. Temperature on a planet, of course, would also affect the amount of hair or fur a creature would have.

STARS AND LIFE ZONES

Life on a planet will also depend on the characteristics of the planetary system's central star and particularly on that star's ability to supply enough heat to maintain water in its liquid state.

Astronomers have tabulated the characteristics of the one hundred stars nearest to the sun. The masses of these stars range from 0.03 to 2.3 times that of the sun. No fewer than eighty-nine of these stars are on the main sequence. (See Chapter Two, "The Curtain Rises.") The main sequence is where most stars spend their lifetimes without excessive flare-ups and without blowing up.

The one hundred nearest stars, presumably typical of stars elsewhere in the universe, include seven white dwarfs and four subdwarfs. Both categories can be written off as to their having life-bearing planets, since the dwarfs of both types are too small and too cold to light the fires of life.

No fewer than seventy-two of the one hundred stars are members of double- or triple-star systems. This, however, does not eliminate such systems as abodes of life, because the members of multiple systems can be far, far apart.

There is still argument in progress whether multiple systems with much-more-closely situated stars could harbor planets with life on them. Some scientists feel that planetary orbits in binary- or triple-star systems would be rather unstable, involving such large variations in the distance from the stars that extreme climatic changes would result.

But we could have easily wound up living in a double-star system. Jupiter came close to igniting the stellar fires. In that case, we would have had two suns, although admittedly Jupiter would have been a small and faint star compared to the sun.

If Mars were somewhat larger and warmed by two suns instead of one, the intriguing possibility of having two life-bearing planets in the solar system would have increased significantly. Life-forms living in double-star systems may have developed the ability to change the color of their skins according to the position of the two suns to derive the maximum benefit from their light.

As it is, single stars would appear to be the most suitable sources of warmth that life needs, and of the nearest one hundred stars perhaps two qualify for that task when their surface temperatures and the types of radiation they emit are considered. When we keep in mind that there are 100 billion stars in the Milky Way galaxy alone, our home galaxy, the total number of stars suitable for life-bearing planets becomes a staggering 2 billion in our galaxy alone. That number represents the estimated number of sun-like stars. (See Chapter Two, "The Curtain Rises.")

Just how wide a zone of life a planetary system has also depends on its central star. In our solar system, it's 1,000 degrees Fahrenheit (538 degrees C) near Mercury and —200 degrees Fahrenheit (—129 degrees C) at Jupiter. Water-and-carbon-based life can exist in only a much narrower zone so that the heat does not vaporize the water and break the carbon bonds, or the extreme cold freeze them.

THE IMPACT OF THE TEMPERATURE

Temperature on and near the surface of a planet is one of the important determinants of the development and nature of living organisms. Human populations find the regions where the mean annual temperature is between 32 degrees Fahrenheit (0 C) and 86 degrees Fahrenheit (30 degrees C) most to their liking. Most of the population lives in that zone, with 95 percent of the world's people crowded into the 65 percent of the land area with a mean annual temperature between 40 degrees Fahrenheit (4.4 degrees C) and 80 degrees Fahrenheit (27 degrees C). What's more, most of the plants and animals important to man as sources of food find that temperature also to be the most suitable.

There are indications from the geological records, preserved in marine sediments, that a relatively warm and equable climate persisted on Earth starting about 65 million years ago. It was followed by a gradual cooling that led to the development of an ice cap on Antarctica 10 to 15 million years ago, and to continental glaciation in the northern hemisphere approximately three million years ago.

Since then, the Earth has experienced oscillations in the climate that have lasted from forty thousand to one hundred thousand years. The ice ages were the most recent stage in climatic evolution. There have been some milder climatic extremes between the glacial and interglacial conditions that have led to a pronounced effect on the early civilizations. But those climatic extremes have not been as severe as the gradual changes that took place during the preceding 50 million years.

The causes of climatic change are also visible in the geologic record. They come about through the movement of continental masses, the uplift and wearing down of mountains, and the changing pathways of ocean currents. Scientists believe that at the present stage of its evolution, the Earth's climate is particularly sensitive to changes in the tilt of its axis and eccentricity of its orbit. Just how sensitive our climate might be to the unintentional changes we are making in our environment, such as massive and continuing release of atmospheric pollutants, is hard to say, because the effect of the introduction of such pollutants may not be felt for hundreds, even thousands, of years. Unfortunately, by that time it may be too late to reverse the damage.

There is no doubt, however, that temperature on a planet is one of the environmental factors that channels the development of animals in specific directions. For many years, for instance, dinosaurs, especially the big ones, were viewed as cold-blooded, slow-moving, and rather stupid beasts. They were said to have been doomed to extinction because of their inferiority to mammals. And because of their supposed cold-bloodedness, the big reptiles were supposed to have lived in an extremely hot climate.

Recently, however, these ideas have been called into question. Champions of warm-blooded dinosaurs have stepped forward to argue that dinosaurs were just as active and speedy as today's large mammals.

The terms "cold-blooded" and "warm-blooded" are not as precise as scientific terms: "ectothermic," or "externally heated," and "endothermic," or "internally heated."

It turns out on closer examination that such an ectothermic, or "cold-blooded," animal as a lizard, which spends its time alternatively running and basking in the sunshine, actually maintains its internal temperature at a level that may be higher than our normal body temperature. The American whip-tailed lizard, for instance, keeps its daytime temperature stable at about 105 degrees Fahrenheit (40.5 degrees C), which would be high fever in man. At night, a lizard's temperature drops and it becomes inactive.

In contrast to lizards and other reptiles, mammals and birds maintain a constant internal temperature regardless of external conditions. A high rate of metabolism does it in mammals and birds, but it demands a higher food intake.

It is now clear that where dinosaurs are concerned, "cold-blooded" and "warm-blooded" may be wrong terms to use.

The largest dinosaurs, for instance, warmed up in the sun and cooled down in the shade relatively more slowly than the smaller reptiles. Even the day-night sequence would have had little effect on a big dinosaur. Accordingly, these beasts must have maintained their internal heat by their own metabolism. They were thus warm-blooded but possibly without a high metabolic rate, and instead of enjoying hot weather they would have suffered in it from overheating. Thus the big dinosaurs could have been warm-blooded even though they didn't gallop all over the place.

More clear-cut is the case of the smaller dinosaurs. The ostrich dinosaurs, which looked uncannily like ostriches but without wings or feathers, are believed to have been warm-blooded. The reasoning goes as follows: the ostrich is one of the fastest-running animals, a performance no cold-blooded creature could match. The ostrich dinosaur is structurally convergent with the real ostrich, which developed much later. Since convergence usually means a similarity in living habits, as we have seen in the placental and marsupial wolves and other convergent creatures described in the preceding chapter, the ostrich dinosaur had to be warm-blooded—or so goes the reasoning.

OUR CLIMATE AND THE MOON

A fascinating, easily overlooked factor in the type of climate we have on Earth is our moon. Aside from its aesthetic and romantic value, the biggest contribution our moon makes to the Earth is to stabilize the orientation of the Earth's axis of rotation.

Mars has no such large satellite and is instead exposed to the combined gravitational influences of the sun and Jupiter. The inclination of the Martian rotation axis, accordingly, has varied over the planet's history from about 0 to 34 degrees, affecting the Martian climate. In contrast, the nearness of our moon has kept things fairly stable on Earth, climatically speaking.

Recently, scientists calculated that the ice ages on Earth could have resulted from very small changes in the tilt of the Earth's axis; the moon apparently made them smaller than they otherwise would have been. Without the moon's stabilizing effect, the Earth could have experienced violent climatic fluctuations on a time scale of hundreds of thousands of years.

Not that a large moon is absolutely necessary to attain the needed stability for life to exist. A planet's tilt, or equatorial inclination, determines the habitability of the various parts of its surface, but the range is rather large. An Earth-sized planet with an equatorial inclination of 75 degrees (the Earth's is 23.5 degrees) would still offer a habitable zone, although this belt would be only two thousand miles wide and would be situated along the equator. In the higher and lower altitudes, annually alternating excessive heat and cold would prevent life from taking hold—at least higher forms of life.

On the other hand, a planet with a low equatorial inclination may boast two habitable belts in the high latitudes.

The Earth could have compensated for an absence of a moon by rotating faster or slower. Venus keeps a stable inclination by rotating very slowly. Similarly, the absence of large tides induced by a big moon probably would not be all that detrimental to the development of life, because small solar tides could compensate for the large lunar ones. (In varied ways, tides may have contributed to the spread of life. When plants began to colonize the Earth's landmasses, for instance, the alternating ebb and flow of water might have speeded the evolution of the root systems in the first colonist plants. So the existence of a large natural satellite such as our moon could be beneficial to evolution and continuation of life.)

LIFE'S INFLUENCE
ON THE ENVIRONMENT

Not only did life evolve under the influence of the environment, but life, in its turn, has exerted a great influence on its surroundings, modifying and changing the environment to make it even more suitable for life.

Microscopic algae, for instance, evolved the wondrous process of photosynthesis, by which the algae, and higher plants, split water into hydrogen and oxygen. They remove carbon dioxide from the atmosphere and combine it with hydrogen to make carbohydrates, or sugars and starches. And plants liberate oxygen, as an unwanted by-product, into the atmosphere. That's what made evolution of higher life possible on Earth and what allows life to continue. Were it not for the oxygen-replenishing ability of plants, the free oxygen in the Earth's atmosphere would be gone in a few thousand years. Oxygen would combine with various substances, while carbon dioxide would accumulate and stifle life.

In this context, the gas giants of the solar system—Jupiter, Saturn, Uranus, and Neptune—are relics of the past, because the developmental processes on those planets didn't go far enough. The gas giants have solid cores, but any development of life on those planets beyond simple microorganisms is being prevented, most likely, by the turbulent circulation of the atmospheres and by the fact that sunlight may not penetrate to those planets' cooler depths. Microorganisms would have trouble evolving in those whirling atmospheres, because on Jupiter, for example, such microbes could stay at a temperature right for life only for a few weeks, or at best months, at a time.

On Earth, plants speeded the development of life in still another way, by establishing themselves as the first land dwellers. The once barren land thus provided food for animals, which could now leave the sea. In more ways than one, in terms of its environment, the Earth is an extremely lucky planet.

OUR SAD SISTER

Venus, on the other hand, once mistakenly called Earth's "sister planet," has been unlucky—showing that life can't arise everywhere. While Venus has a thick atmosphere, made up almost entirely of carbon dioxide, there is very little carbon dioxide in our atmosphere. On Earth, carbon dioxide is locked up in limestone rocks, and these rocks, in turn, consist of an agglomeration of tiny seashells that were constructed by sea creatures from the carbon dioxide dissolved in water. Like photosynthesis by green plants, this is a continuing activity by those sea creatures, keeping the amount of carbon dioxide on Earth under control.

On Venus, no such controls existed, and carbon dioxide came to dominate the planet's atmosphere.

Venus still presents a large number of puzzles. Radar found it to be rotating in a retrograde manner, that is, in a direction opposite to its revolution about the sun, and at that rotating only once every 243 days, 2 hours, and 40 minutes. Some sister! Except for the marginal case of Uranus, which like a tipped-over child's top lies in its orbit on its side, Venus is the only planet that rotates backward. Just why this is so, no one knows for certain, although scientists theorize that Venus's extremely heavy atmosphere might be a factor.

In addition, when Venus and Earth are at their closest, Venus always faces the Earth the same way. The Earth apparently exerts a grip on Venus—a kind of interplanetary handclasp. Theoreticians predicted that the Earth's pull would produce a permanent bulge on Venus a few thousand feet high, and later such a bulge was indeed detected by radar, in the right spot.

But Venus would need an internal mechanism to dissipate tidal energy. Scientists think that only a liquid core could provide this outlet. Atmospheric tides, the permanent deformation of the equator, and a liquid core apparently contrive to place Venus in its strange state of motion.

Another riddle is what happened to all the water that scientists assume must have been abundant on Venus. They assume the initial presence of water on all the terrestrial planets, and on the moon, too.

The terrestrial planets are believed to have generated water from their interiors along with the gases that seeped out. Water may be locked under a layer of permafrost on Mars and on the moon even today. On Venus, there is so little water in the atmosphere that if it could be evenly spread over the planet's surface, it would constitute a cover only about one eighth of an inch thick. In a dramatic contrast, if similarly spread out, the Earth's oceans would cover our globe to a depth of nearly two miles.

Being about one third closer to the sun than the Earth, Venus could have started out with less water, of course, or its oceans could have been destroyed by ultraviolet radiation from the sun, which can break up water molecules. In any event, scientists find it much easier to explain Venus's carbon-dioxide atmosphere than the absence of water. Were the temperature of the Earth raised to that of Venus, the Earth, too, would have had a predominantly carbon-dioxide atmosphere. Locked up in the Earth's crust is about twenty-five atmospheres of CO_2 as "fossil gas" in the form of limestone and other minerals. As we have seen, on Earth, life and water made the difference.

Carbon dioxide at the same time accounts for much of Venus's heat. CO_2 lets in radiation from the sun but blocks infrared heat rising from the surface, thus creating a "greenhouse" effect. The glass in a greenhouse, playing the same role as the CO_2 in the atmosphere of Venus, lets visible light in but traps the heat rising from the soil. A small amount of water vapor in the atmosphere makes the greenhouse even more efficient. The surface of Venus is obviously too hot for carbon-based life—a striking example of how a planet's environment makes a difference between life and its absence.

As a result of scrutiny by spacecraft, Venus came out looking stranger and more inhospitable than anyone ever imagined: an awesome oven of a planet, where sunlight may not reach the surface, which may be glowing red in spots. Extrapolating from Soviet and U.S. data, scientists have put the surface pressure on Venus at 100 atmospheres and the surface temperatures at 900 degrees Fahrenheit (480 degrees C). Pressure at sea level on Earth equals one atmosphere.

THE MYSTERIES OF MARS

Mars might have been luckier than Venus. It might have put a toe through the gate of life—just. Although, like Venus, it now has an atmosphere dominated by carbon dioxide to an extent of 95 percent, it's a very tenuous atmosphere, because Mars is a smaller planet. Most of the carbon dioxide at the planet's surface, furthermore, appears to be locked in carbonate rocks and in the polar caps. Scientists have calculated that if all that carbon dioxide were liberated, the atmospheric pressure on Mars would rise to about half its value on Earth.

This means that liquid water could have existed on Mars—a conclusion supported by spacecraft photographs of those tantalizing dry riverbeds on Mars. But careful study of the photographs reveals no dry lakes or any other signs of large water bodies or slow-flowing rivers.

What the pictures seem to show instead are riverbeds carved by fast-flowing water, much as is done by flash floods in the deserts and semiarid regions on Earth. Some of the riverbeds originate in depressed areas of jumbled hills, indicating a collapse of underground reservoirs and sudden release of water. Some of these beds run for hundreds of miles and are as wide as six miles. These dry riverbeds, incidentally, have nothing to do with Martian "canals," which have turned out to be canyons or streaky dust deposits.

So if water existed on Mars, it's still unknown whether it existed long enough for life to have evolved. Outlines of impact craters in the floors of the channels show these craters to be about three billion years old.

"The evidence is reasonably convincing that there is no contemporary life on Mars," says Dr. Richard S. Young, chief of biology at NASA. "But Viking tells us nothing, or very little, about the history of the planet, and we don't know what happened ten thousand, a million, 10 million, 100 million, a billion years ago. We have no idea whether life appeared on Mars or not. Whether life got started and didn't make the grade. Or whether life got started and just barely made the grade and is hiding out twelve feet below the surface, or inside rocks. I think there is a multitude of clues left that we were simply not able to get at. Viking just scratched the surface, literally and figuratively. Returning a soil sample from Mars would help."

As we have seen, among the many distinguishing characteristics of the environment that have a significant bearing on a planet's ability to support life is its distance from its sun. The zone of life in our solar system, for instance, appears a lot narrower than once supposed. If we put the Earth closer to the sun, where Venus is, the Earth would become like Venus. Oceans would evaporate, producing that famous greenhouse effect and heating up the planet to a degree that would make life unbearable.

If we placed the Earth where Mars is, on the other hand, our oceans would freeze. Or at least some scientists think so. Some intermediate possibilities have been suggested, such as a halfway-station kind of climate—a warm zone around the equator only, but the atmosphere consisting of carbon dioxide, suitable for primitive but not advanced life.

Being only little more than half the diameter of the Earth, Mars apparently could not hold on to its internal heat. Yet volcanoes spewed gases into the Martian atmosphere until as recently as 100 million years ago, when dinosaurs ruled the Earth. Mars appears to have possessed a dense atmosphere at one time, as well as water—those intriguing dry riverbeds are proof of its existence. Life may have started on Mars and then died out after the volcanoes fell silent. At that time, the surface water and the Martian atmosphere, no longer fed by the geophysical heat machine, would have dissipated into space.

A planet's size and bulk, suggests Robert Jastrow, may be more important than its exact place in its solar system, because size and bulk determine whether a planet will have oceans, which vastly enhance the possibility of life developing.

If Jastrow is correct, had Mars been bigger and had its atmosphere evolved in a fashion suitable for life, it is entirely possible that we would have had two planets with life on them in the same solar system—with all the exciting possibilities that implies.

Such findings and controversies have inspired Earth-based scientists. They feel that even if ultimately no life is found on Mars or elsewhere in the solar system, exploration of our solar area will be one of man's most exciting and productive enterprises.

"One way or another, there's bound to be something of volatile interest, even if the planets turn out to be lifeless," says Carl Sagan. "There is a continuum of possibilities leading up to and beyond the origin of life on a planet. You might have had a climate in which only prebiological organic matter was produced. Organic matter was produced on the Earth before there was life—obviously that stuff was needed to make life. If there was a place where life had never come into being but organic matter was lying around, then we have the building blocks of life lying around. We could tell something about the origin of life.

"Or it's possible that life arose on a planet and subsequently became extinct and so we can look for fossil and other evidences. It's possible that life arose on a planet and didn't evolve very far but is still hanging around. There is a whole spectrum of possibilities, but even in the case of a lifeless planet, there's a possibility of earlier life, or the possibility that life never arose but that there is prebiologic organic matter. Even such unlikely objects as meteorites, which probably come from asteroids, which have no air and no water, still are loaded with organic molecules that come from their parent bodies. So we surely should find out some things about the origin of life even if we study a lifeless planet."

THE IMPORTANCE
OF CONTINENTAL DRIFT

To squeeze through the gate of life, and to produce intelligent life, still another environmental influence had to be at work. It appears that to harbor life a planet must also have undergone geological processes similar to those that took place on Earth. Specifically, these are plate tectonics and continental drift.

Plate tectonics—the word means construction—refers to the buildup of the Earth's geology by the action of plates on which continents and oceans ride. These plates constitute the outer shell of the Earth's surface, a few miles thick. Heat from the Earth's molten core and decay of radioactive elements keep the plates in motion, bumping into one another. Where the plates meet, earthquakes occur and a restructuring of geological features takes place. Volcanoes mark the boundaries of the plates; the Earth has about ten large plates and about thirty smaller ones.

Astronomical observations and investigations with spacecraft of planets and large satellites in the solar system show that plate tectonics and continental drift are not wholly universal processes. Still, they are not totally confined to Earth, either.

Perhaps not surprisingly, being small, the moon and Mercury show no tectonic activity. There are faults and dead volcanoes on Mars, indicating there might have been geologic restructuring in the distant past. Today, though, the red planet appears quiescent.

Venus, unlike the Earth, appears to be a one-plate planet. About 85 percent of the Venusian surface may be composed of a giant, planet-wide continent made of light-weight granitic rock that was extruded early in the history of Venus. The Earth's "sister planet" appears to have had higher temperatures in its interior early in its history due to an absence or loss of water. This higher heat would have led to more efficient settling out of heavy basaltic rock into the planet's lower crust. The lighter granitic rocks would then have risen to the surface.

Finally, instead of ocean basins that cover five sixths of the Earth's surface, Venus has only a small amount of low-level terrain that faintly resembles ocean basins. If there was tectonic activity on Venus, it appears to have been choked off early on.

One body in our solar system that *is* geologically active is Jupiter's moon Io. In the spring of 1979, the American spacecraft Voyager 1 flew by Jupiter and returned some spectacular views of the volcanic eruptions on Io. No fewer than seven volcanoes were going simultaneously. Although different from the terrestrial volcanoes because, being smaller than the Earth, Io long ago lost whatever water it might have had, these are volcanoes nonetheless, spewing gases nearly 100 miles (160 kilometers) high, with an ejection velocity of 1,200 miles (2,000 kilometers) an hour.

Volcanic activity on Io means that geological upheavals are in progress there, such as motion of surface rocks and friction of rocks beneath the surface. Io is not big enough to have a molten core and radioactive materials in its bowels—it's not big enough to have retained those sources of heat. The friction of Io's rocks, instead, is caused by Jupiter's gravitational pull acting as a tide.

Even though the source of energy that causes plate tectonics on Io is different from ours, the fact that the process has been shown to be at work on two bodies in our solar system demonstrates that the Earth is not unique in that respect.

This is important to know, because plate tectonics and continental drift could be vital to the emergence of land-based intelligent life.

To start with, the breakup of the supercontinent, Pangaea, created a large number of continents and islands, thus offering a great variety of ecological settings for life to exploit. As continents collided, came apart, and collided again, new settings were created, new species emerged, and differing kinds of life intermingled. Into the openings created by land bridges poured hordes of immigrant animals: the invasion of South America by North American placental mammals is a prime example. (See Chapter Four, "The Turns and Twists of Evolution on Earth.") Without such large-scale drifting and collisions of continents, the course of evolution could have been decidedly different.

Without plate tectonics and continental drift, for instance, the whole of the Earth would have been covered by one vast ocean. In fact, a gigantic ocean without end once covered the whole of the Earth; then plate tectonics lifted up some of the crust and made it into continents. As we have seen earlier in this chapter, most of the surface of Venus, on the other hand, where water is almost nonexistent, constitutes a giant, planet-wide continent—which would have been the Earth's fate too, were it not richly endowed with water. While development of life is possible on an all-ocean planet—life on Earth developed in the ocean—life on a world that remains an all-ocean planet will be a more limited kind of life than we have on our land masses today. (See Chapter Six, "Ocean Kingdoms and Insect Worlds.")

Plate tectonics and continental drift also forged the materials that got civilization on Earth going. It was at the converging plate boundaries that chemicals got melted, recrystallized, and pressurized to be converted into metallic ores. Where continental plates meet often lie significant ore deposits. Such deposits are found at both past and present plate boundaries. In Cyprus, for instance, where the African plate nudges the European plate, occurs copper of high purity. Lead and silver are found across the western United States, the border line of the East Pacific plate.

The thick deposits of coal were created by colliding continents when they were lying in the tropics. The tropical jungles on the shores of the continents became slowly submerged; layers of vegetation were covered with water before they had a chance to rot. New forests sprung up on top and were in turn drowned. The land sank and was covered by the edge of an overriding continental plate, squeezing water and gases out of the peat-like mass. Thick deposits of coal were the result.

Similarly, oil deposits were created as the sea floor crumbled in the collision of continents. Trapped below were remains of marine plants and animals. Bacteria attacked these remains, removing oxygen from the carbon molecules and thus creating the hydrocarbons.

Improved soil for farming also owes its origin to the violent geological processes. Volcanic rock, for instance, offers a greater variety of plant nutrients than does granitic rock, because volcanic rock weathers more readily.

Uncannily, deposits of metal ores, fossil fuel, as well as rich farmland happened to be concentrated to a large extent in the northern hemisphere—conveniently situated resources for a developing civilization to tap.

Such remarkable interplay between the emergence and development of life and physical and chemical processes on a planet is bound to take other forms elsewhere, of course, but its importance can hardly be doubted.

IS EARTH ALIVE AS A PLANET?

In fact, scientists have lately evolved a concept of life on Earth being integrated on a global scale. What they are saying is that the Earth is alive as a planet. The concept has been named after Gaia, the earth goddess of the Greeks, and it has been developed by American scientists James E. Lovelock and Lynn Margulis.

Their argument is that life responds and reacts to outside changes by modifying conditions on Earth to make them suitable for the continuation of life. As our aging sun increases its temperature and sends more heat toward the Earth, for instance, life responds by modifying the Earth's atmosphere and surface geology to keep the climate fairly constant. Another example of Gaia at work would be the maintenance of the oxygen content of the Earth's atmosphere at precisely 21 percent; a larger percentage would have increased the possibility of forest fires set off by lightning to an enormously dangerous degree. Life, in short, according to this view, is composed of a single fabric—a sobering thought but not an outrageous one.

The existence of Gaia can be proved only indirectly. But there is no question that in many highly visible ways the Earth and life on it are interrelated.

Adaptation of animals to their environment illustrates this point well. Adaptation is one of the main themes of evolution; the other is diversity of living things. The two are opposite sides of the same coin, because both stem from life's successful fitting of its environment.

Both adaptation and diversity take place when a sudden opening in the environment draws a new group of organisms into a new way of life.

As they search for food and shelter, living things begin to specialize narrowly, as woodpeckers and frogs do, for instance. The adjustment to a new setting can make differing specialists out of a single species, as Darwin's finches on the Galápagos Islands demonstrate so beautifully. Surviving in a new environment forces still other creatures to change their lives to entirely new ways, as dolphins and whales did when they abandoned the land for the sea.

"Life," wrote the noted American geneticist Theodosius Dobzhansky, "tends to spread out and utilize every opportunity for living, no matter how narrow and constraining it may seem to us."

During its expansion, each form of life is sculptured in part by its genetic possibilities and limitations, and in part by the environment in which it and its ancestors arose. Indeed, the bodies of animals on Earth are composed largely of the medium in which ancestral forms of those animals lived: the seas. The salinity of our blood, close to that of the early oceans, is a reminder where we came from.

The reason we have such a fantastic variety of life on Earth today is that animal life has tried to solve new problems in an ever-changing environment in new ways. The dictate of evolution is simple: adjust to change, or die. Interestingly, animals that had found ways to successfully cope with environmental challenges by selecting narrow niches where predators are few, have persisted in forms that date back millions of years. Frogs are one current example.

Relationships among different creatures, in fact, are an important factor in animal evolution aside from the genetic and environmental components. These relationships can influence animal development in decisive ways.

After predators first appeared in large numbers in the seas, millions of years ago, for instance, all kinds of marine creatures suddenly began to grow protective shells and outer skeletons. After amphibians multiplied on land, insects took to the air to escape from them. After giant reptiles invaded the seas, sharks had to become more efficient and more ferocious to protect themselves.

A changing environment alters a life-form. Typically, the horse started out as a small forest dweller, no bigger than a cat, feeding on soft shoots. But as forests declined in area, the horse emerged into the prairie, feeding on tough grasses and growing bigger to successfully get away from predators.

Of course, as you've seen by now, there are many ways to adapt to the environment. To secure oxygen from different environments, terrestrial animals evolved lungs while aquatic animals use gills. To survive in changing environments, many creatures blend into seasonal variations by changing their colors. The snowshoe rabbit, for example, is brown in the summer but grows white fur in the winter. The flounder can change its color even faster. When on a sandy bottom, it displays small spots; but on a stony background these spots grow in size to match the size of the stones.

Adaptation does not necessarily lead to perfection, or at least to an ability to cope with unforeseen events.

Species become extinct, but students of evolution don't find the fact of extinction *per se* surprising. The fossil record reveals that extinctions are the rule, but the rates of extinction vary significantly through time.

At one time, some scientists theorized that lineages of animals had life cycles much like individuals, proceeding from youth to maturity to old age and death. The concept of "racial aging" was introduced.

More recently, however, this whole idea has been shown to be without substance. The changes those scientists had misinterpreted as signs of senescence were really adaptive. It's changing conditions in the environment, rather than any innate aging of a species, that causes its demise.

Events external to the planet can also influence life on it, of course. This may have happened on Earth. It's been suggested, for instance, that an exploding star triggered changes on Earth that led to the death of the dinosaurs and many other forms of life at the same time the dinosaurs were dying out. Another suggested reason for the death of the marine creatures alone is an overflow of the Arctic Ocean, which in ancient times may have been a gigantic freshwater lake. The rush of fresh water would have greatly reduced the ocean's salinity, killing off species that could not adapt.

No single theory suggested so far, though, explains why so many species—about half of all species living then—vanished. While we can't rule out "a banana peel on the last step, a final cataclysm that knocked the dinosaurs off," as Dr. William Clemons, a professor of paleontology at the University of California's Berkeley campus, puts it, there appear to have been gradual changes taking place over millions of years toward the end of the Age of Reptiles that affected various animals in various ways.

What it all adds up to, as Dobzhansky and his associates have pointed out, is that evolution has no foresight. Evolution succeeds in adapting species to existing conditions, but when such conditions change too rapidly for a species to adjust, the species perishes. Something changed in the environment, but dinosaurs apparently failed to keep pace. It is erroneous, though, to say that dinosaurs were nature's "mistake"—they ruled the Earth, after all, for far, far longer than man has. Had they lived on, perhaps man would never have had a chance to arise—only intelligent reptiles.

Natural selection thus offers an explanation for both the fantastic diversity of life on Earth and the disappearance of species. There is no reason to believe that creatures on other planets will not similarly exploit all possible environments—in the air, on land, and in the seas. Similarly, no species anywhere is going to be immortal.

Life here is part of the spectrum of life everywhere. Chance obviously plays a role in evolution; if the cause of the dinosaurs' demise was, indeed, a chance event, man may be here by chance. Yet the history of evolution suggests that adaptation is the nonrandom element in evolution. What this means is that there are certain predictable ways, as we see on Earth, in which animals can adapt to their environment. The pressure for change results from a combination of physical, genetic, and ecological opportunities. They control the direction of evolution but don't limit it to a single direction—allowing us to expect the rich diversity of paths that life has taken in Darwin's universe. As Carl Sagan explains it: "We have a worrisome tendency to think that what we see is all that can be. But it seems very clear that if even minor events had gone slightly differently billions of years ago the organisms today would also be very different; this is particularly true on a molecular biological level. Of course convergent evolution occurs when there is one best solution to a given physical problem as, for example, imaging light at optical frequencies or high-speed transport in water. But why five fingers? Why fingers rather rather than tentacles? Why the agonizingly slow data processing in our neurological systems? Why not multispectral infrared sensing? It's easy to think of a wide range of anatomies, physiologies, and sensory modalities that have not been adopted by humans or indeed by any other creatures on the Earth." Some of these possibilities are explored in the next chapter.

CHAPTER SIX

Ocean Kingdoms and Insect Worlds

In a universe of an unimaginable diversity of life, the road to higher intelligence would be open to a great variety of creatures. Even on Earth, where man prides himself on being the ruler of the planet, the possibilities for other creatures to assume that role were always there. Such opportunities may well have been seized on other worlds by beings as diverse as an octopus (or its likeness), by insects, or even birds. For all creatures, the signs point toward the emergence of intelligence, even in the lowliest of beings. "If a slug can give rise to an octopus," says Dr. Berrill, "then all kinds of things are possible. Any form of life under certain circumstances could conceivably become intelligent."

A world as unlikely as—and yet in many ways like—the one we inhabit would be a planet completely covered by a single, gigantic ocean. Such an eventuality, as we have seen in the preceding chapter, can take place on a planet with a somewhat different, less violent geological history than ours where continents never quite break through to the surface. It would probably be a larger planet than the Earth, because a larger planet would push out relatively more water from its interior.

In fact, a huge ocean covered the Earth's surface until about four billion years ago. Then came a cosmic cannonade: the intense bombardment of the Earth, the moon, and other "terrestrial" planets by huge asteroids and meteorites, which may have been the remnants of a planet between Mars and Jupiter that broke up, leaving the asteroid belt behind. The impact of these cosmic missiles produced huge chains of craters, with many of the individual craters as wide as six hundred miles. These indentations became the original ocean basins on Earth.

The formation of the impact basins wreaked havoc with the Earth's geology. It changed the underwater topography in a dramatic manner, since the original impact basins were as deep as ten miles. It triggered a partial melting of the underwater crust, leading to its extensive fracturing and the upwelling of the basaltic lavas. Finally, the formation of the impact basins enhanced the temperature variations in the crust, encouraging the formation of rift valleys and mountain building. The continents began to surface.

Still, the greater part of the continental surfaces continued to be covered by the so-called epicontinental seas, which were several hundred feet in depth. Sufficiently illuminated by sunlight and with the water layer providing protection against damaging ultraviolet radiation, these platforms became the stage for the development of life. The shallow seas were the original birthplace of life and its home for two billion years. Here evolved a remarkable range of creatures: sponges, jellyfish, sea anemones, snails, mussels, squid, crabs, lobsters, sea urchins, starfishes, octopuses, and many others. Marine creatures were able to develop solid bodies with protective shells, since there was no need for them to display a density greater than that of the surrounding sea water, something that would have been required if they had developed at greater depths. Here they had the continental shelves as convenient "floors" to support their bodies.

To this day, many marine organisms prefer to inhabit the shallow continental shelves, rather than the ocean deeps. Shallow water may also be desirable for the chemical reactions that created the first organisms on Earth—such depths, to be sure, can exist without continents surfacing.

From the sea life on these submerged platforms descended such denizens of the deep as the giant squid and other creatures found in the ocean depths.

In the sea evolved those two radically different shapes of living things: the radial form of the starfish and the bilateral shape of the squid.

Some of those creatures of the sea—anemones, jellyfish, starfish—had beautiful, symmetrical bodies, but they lacked intelligence.

Others were incredibly ugly, even frightening. The *Vampyro-*

teuthis infernalis, "the infernal vampire squid," with its mask-like body and ten tentacles extending from its lower portion, survives to this day. It looks as sinister as any creature ever imagined by any science fiction writer. (See illustration page 144 in Chapter Seven, "Life Beyond Darwin.") Fortunately, the sole surviving variety of this creature doesn't grow longer than fourteen inches and lives mostly out of sight of man, at depths of eighteen hundred to ten thousand feet. Were it bigger and more visible, the vampire squid surely would have qualified for the Hall of Fame of Monsters.

Even on Earth, where the seas now cover not the whole planet but 71 percent of it, a remarkable 150,000 of our planet's nearly 2 million species are found in the oceans.

The richness and the color of life in our present-day oceans can be only hinted at. It bursts on an underwater observer like a fairyland of a thousand glittering Christmas trees, with the most imaginatively wrapped presents spread around them.

It's a world that in many ways is still poorly known, in whose depths strange creatures light up the darkness with fantastic organs of bioluminescence; where giant squid, whose occasional appearances on the surface of the sea gave rise to the sea-monster myths, battle giant whales; where ocean currents big and swift enough to match our mightiest rivers flow unseen.

One big difference between the ocean and the land is that, in the sea, the environmental conditions are fairly stable. This would encourage marine organisms to become widespread, occurring almost everywhere in the sea.

If continents had never surfaced, if Earth or another planet were totally covered with water, what kind of life would such a planet boast? And how far in intelligence could such life-forms proceed?

First, we would have to subtract from such marine population the mammals that went back to the sea: dolphins, whales, walruses, seals, sea cows, and a number of others.

The seas' indigenous population would then consist of two major kinds of animals that are exclusively marine. First, there are the mollusks, such as octopuses, squid, clams, and oysters. Second, there are the spiny echinoderms, which include starfish, sea urchins, and others. There would be no fish of the kind we know, for vertebrate fish are the product of freshwater streams. Some of them later migrated to the sea to give rise to such cartilaginous fishes as sharks. So even sharks are out, perhaps to make for a more peaceful world.

Instead, on a planet entirely covered with water, we would find strange-looking fish-like organisms like those shown in the color insert.

The fish-like creatures would have evolved flexible vertebral rods as support for their body muscles. They would have a siphon for propulsion, of the type used by the octopus and the squid, and would also push themselves along by a wave-like movement of their fin-like bodies, much as a water snake does. Because this "fish" would have evolved from a slug, it wouldn't have the type of fins with which we are familiar.

Still, life on an all-water planet would look familiar to us, because there would be a high likelihood of an extension of evolution along a kind of molluscan line—squid, octopus, and so on. They would be large and small and somewhat different from those in our oceans, but not so outlandishly so. That similarity of form, of course, is predicated by the physical constraints that force life into certain practical channels. The fast swimmers would have fish-like shapes, while slower-moving creatures like starfishes and octopuses would display a radial symmetry.

There will be differences, of course. On a large planet covered with water, because of greater gravity, the creatures would be less dense in structure. The pelagic drifters would need more body surface, either as their filament or as rod-like projections, and bottom-walkers would need sturdier legs. Swimmers could increase their surface relative to volume by becoming more elongated, that is, more snake-like. This shape tends to reduce head-on resistance and at the same time exploit the stickiness of water in reducing the rate of sinking. This would apply to any creature with tissues and skeleton heavier than water. Cephalopods on Earth in great depths have apparently overcome this difficulty by becoming the same density as water and are translucent and widely stretched out—ghosts of the deep.

Of all the marine beings, an octopus-like creature would have the best chance to emerge as the ruler of an ocean kingdom, becoming the kind of superoctopus shown in our illustration. This king of the seas could have two telescopic eyes looking upward to the light and surface of the sea. Around the middle of its head, it could have a ring of large camera eyes. It could have as many as twenty or forty tentacle arms, each subdivided into three tapering, flexible fingers. Tentacles as long as one hundred feet would be needed to catch crabs and

other delicacies. When coming close to each other, the giant octopuses would communicate by rapidly changing their colors.

Even on Earth, where the octopus is still a relatively little-known
creature, it is beginning to acquire a reputation for high intelligence
and a wide range of emotions, despite its walnut-sized brain.

The octopus was much maligned in the past, and both myth and
fiction described it as a sinister monster. But undersea explorers such
as Capt. Jacques-Yves Cousteau found the octopus to be a shy, retir-
ing creature that stays out of sight in its hole under large rocks. Octopus "houses" take on many forms: large rocks, or piles of various
materials: stones and bits of metal, decorated with shells.

In his book *Octopus and Squid* (Doubleday), which he coau-
thored with Philippe Diolé, Cousteau recalled fondly his discovery of
"Octopus City," off the Mediterranean island of Porquerolles. Cousteau and his associates had spotted a series of small hills spread out
over a flat stretch of the sea bottom. In each hill—an octopus "house"
—the explorers found an octopus living in its own little cottage.

Feeling like intruders, the explorers could make out the globular,
black-pupiled eyes of the octopuses, with their alien, horizontal-bar
pupils, eyes that stared back at the divers just as curiously as the
explorers did at the strange householders—an encounter between species from different worlds.

It appears that octopuses build such houses when caves, which
they often favor as homes, are absent. The octopuses display considerable ingenuity in building their houses. Cousteau and his friends,
for instance, were astonished to see one octopus living under a tree
trunk that it had raised and buttressed with bricks. (An extraterrestrial octopus-like creature might engage in even more complex construction, as shown in our illustration.)

"Was it a form of intelligence?" Cousteau asked. "I still believe it
was."

Divers encountering octopuses at close range noticed that the creatures would protect their ramparts with an outstretched arm, changing color and keeping their eyes on the human intruders. The creatures would also pile up their pebbles and shells as if to try to bar the
invaders' way. The octopuses struck such observers as Cousteau as
displaying "a psychic makeup far more advanced than we had
thought."

Furthermore, explorers have found that, far from trying to devour

them, the octopuses after a while would even allow the divers to play with them as if they were pets.

Rising toward the surface, with its head shaped like that of a wise old man and its arms covered by its mantle extending downward, an octopus looks like an apparition from outer space, a fitting occupant of a flying saucer—a being as alien as any on Earth.

Alien, and yet, despite its separate evolution, equipped with organs we readily recognize: a heart, camera-type eyes, stomach, gills to breathe oxygen. All this despite the fact that the common ancestry of cephalopods and mammals is extremely remote, perhaps as remote as the amoeba.

It is true that mammals descended from fishes, which already had such organs. But parallel evolution of such organs took place in invertebrates and insects, too, as we can see in the comparative illustration of a cat, a grasshopper, and an octopus, on the following pages.

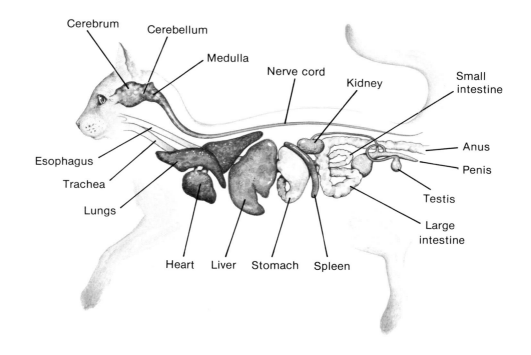

Cerebrum Cerebellum

Medulla

Nerve cord

Kidney

Small
intestine

Anus

Penis

Testis

Large
intestine

Esophagus

Trachea

Lungs

Heart Liver Stomach Spleen

*Comparative anatomies of cat,
grasshopper, and octopus.*

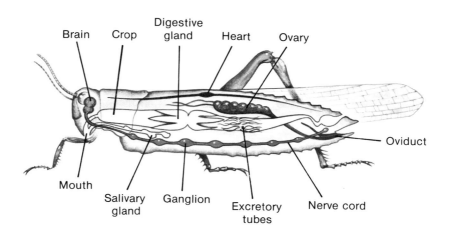

Brain Crop Digestive
gland Heart Ovary

Oviduct

Mouth

Salivary
gland Ganglion Excretory
tubes Nerve cord

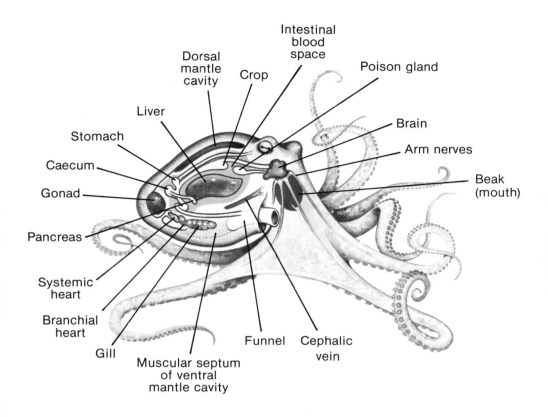

Intestinal blood space

Dorsal mantle cavity

Crop

Poison gland

Liver

Brain

Stomach

Arm nerves

Caecum

Beak (mouth)

Gonad

Pancreas

Systemic heart

Branchial heart

Gill

Funnel

Cephalic vein

Muscular septum of ventral mantle cavity

The positions of these organs are not always where we would expect to find them. The nervous systems of insects, for instance, run alongside the underside of their bodies, their hearts are on top and to the rear of their digestive tracts, and their organs of hearing are often found in such bizarre places as the abdomen, as in grasshoppers and moths, and on the front legs, as in crickets.

Naturally, there are also important differences between the inhabitants of the sea and the land dwellers. Mollusks such as the octopus, and crustaceans, too, for instance, have a copper-based respiratory pigment, hemocyanin, in their blood, which gives it a blue-green or bluish color. Man, all mammals, and most fishes, on the other hand, are equipped with the more efficient, iron-based respiratory pigment hemoglobin, which makes up the coloring matter of blood cells.

Insect blood comes in a variety of colors, often yellow or green, but almost never red. It isn't red in land-based insects, because their specific method of respiration, through air tubes through which oxygen reaches the tissues directly, makes the use of a respiratory pigment unnecessary. But the blood of the larvae of certain aquatic insect species does contain hemoglobin; these larvae have gills instead of air tubes and need the oxygen-carrying pigment to breathe underwater.

Once again, though, it's not the differing shapes of internal and external organs and their unusual locations, or the different types of blood, that distinguish these different animals. What is impressive, rather, is the fact that such parallel organs and tissues exist in almost totally unrelated animals, stressing again the universal nature of biology. Although these organs may differ in design, their function is clearly the same. In the sea, on land, or in the air, the basic scheme of life is one. This means that creatures on other planets will have arrived at similar solutions as well. So a water planet would contain life even more similar to our sea life than would be land-based life elsewhere compared to our land-dwelling animals.

An octopus-like creature elsewhere is bound to have a high aptitude for learning and remembering. Our octopus is a distant cousin of the oyster and the clam, but in its intelligence it far outshines those sluggish creatures. It can distinguish size of objects, choose between complicated patterns composed of lines and designs, make detours through an experimental maze to reach a crab it has seen through a glass wall, and open a glass jar to get a lobster placed inside by experimenters.

The octopus' means of locomotion sets it apart from many other inhabitants of the sea. To get away from its enemies and to move about in open water, the octopus employs jet propulsion. It inflates its mantle and expels water through a tube-like funnel. To hunt crabs, however, octopuses "walk," in a strange flowing motion, on the sea bottom.

The octopus, furthermore, boasts highly developed organs of sense. Those camera-type eyes of the octopus, for instance, are almost as good as human eyes. The creature can change its appearance to an amazing degree, both in shape and in color, to camouflage itself and to express its emotions. On top of that, the octopus has arms perhaps even more dexterous than human hands, with its suction cups giving the octopus a sensitivity of touch unknown to humans. All in all, the octopus is a highly promising candidate to evolve under appropriate conditions into a creature much more intelligent than it already is.

The science fiction writer Murray Leinster gave us a preview of such a creature, a kind of sage of the sea. In his short story "De Profundis," Leinster sketches a mysterious intelligent creature that lives in the water abyss below, without knowing anything about what lies above the surface. The creature appears to be a combination octopus-squid-scallop, with long tentacles and the scallop's eighty eyes. The huge creature and its likes communicate with each other by telepathy—but with an interesting twist. Being highly cannibalistic, the creatures have to maintain a mental block about the location of their homes, lest they be devoured by their compatriots.

Leinster might have had the right idea, at least about the creatures' size. The biggest octopuses encountered by divers so far in our oceans are about thirty feet across and weigh up to two hundred fifty pounds.

But much bigger octopuses inhabit the ocean depths. Although such giants have never been captured alive, their remains are found from time to time. The biggest cadaver of an octopus, weighing an astounding six tons, was found in 1897 at a beach at St. Augustine, Florida. When alive, this specimen was about 25 feet across, with arms 75 to 90 feet long—an animal whose total length was more than 180 feet.

Given nearly ideal evolutionary conditions, such as abundant inorganic nutrients, right temperatures, and extensive shallow seas, says the Canadian paleobiologist Dale A. Russell, it wouldn't be surprising if octopus-like creatures evolved man-sized brains and a capacity to manipulate objects even more skillfully than they do now. Such an evolutionary development could have happened in about 500 million years, Dr. Russell estimates, faster than the 700 million years that elapsed on Earth between the appearance of the multicellular metazoans and *Homo sapiens*. "These creatures," says Russell about the intelligent octopuses, "could be expected to have relatively high metabolic rates, low reproductive rates, and long life-spans, much of which would be invested in caring for their young."

There is, to be sure, not much future in a sedentary life in the black abyss of the sea or even in shallow seas. Life on land provides far more opportunities for the development of a creature's intellectual and physical potentialities. So let's take the octopus out of the water. Imagine a planet somewhat like Earth, with continents and vegetation but without other land-based life, except perhaps for insects. If we transform the octopus into a landlubber by giving it a backbone, integument impervious to water or controllable as in insects, and lungs, it could then emerge on land without committing suicide and become a formidable creature, intellectually and physically.

"Suppose ancient marine mollusks had an outer, flexible shell that was truly part of their body, like the outer integument of insects but more flexible," says Dr. Berrill. "Their bodies would retain water and salts in the manner of fish and other vertebrates. As it is, if you take an octopus or squid or sea slug out of the sea, it dies because it dries up. Yet for a while an octopus out of water can clamber about with the aid of its arms very effectively. If their ancestors were so equipped to resist desiccation, they might well have made a direct migration from the sea to the land, without having to take the freshwater route that fishes did. If it could survive out of water, the sky would have been the limit."

The road toward the sky may lead through the octopus arms. In the octopus that lives in our oceans, not all the arms are identical. Its two arms in the axis of the eyes, the dorsal arms, are exploratory and prehensile. The octopus uses these arms to feel and grasp. It employs the next two arms to pick up pebbles to build ramparts around its house and to catch crabs, its favorite food. The ventral arms are used by the octopus to keep anchored to its rock or its house. All in all, the octopus displays great coordination of its movements; it can, for example, seize a crab with one arm while warding off an undesirable object with another.

In a land-dwelling octopus, such capabilities could be significantly enhanced. Such a creature could walk on four arms, which would turn into legs, and have four additional arms free for manipulating objects. Already displaying a prodigious talent as a builder, a land-dwelling octopus could become a real architect.

The brain of a land-dwelling octopus could become extremely large. In the sea, there isn't as much diversity to challenge an animal; the dolphins and other mammals that went back to the sea are an exception where brain size is concerned. They probably had good-sized brains when they went to the sea. In any event, a dolphin's brain is often larger than man's, while that of the octopus is no bigger than that of a squirrel.

With its arena of stimulation widened spectacularly, a land-dwelling octopus could develop a large brain and engage in activities that would be unthinkable in the sea. Aside from becoming a skilled builder, it might master fire-making and even start a technological civilization. But such a leap is not entirely necessary for an intelligent creature. The land-dwelling octopus, therefore, might choose not to advance higher technologically than, say, becoming a highly skilled gardener. Lack of technological prowess, of course, is not a mark against a creature's intelligence. It might actually be a big plus where its continued survival as a species is concerned, since its means of destruction would be limited. (See Chapter Ten, "Contact!")

Another group of creatures whose ancestors came from the sea—insects—could have had a clear express lane to intelligent evolution if no other life-forms competed against them. Such a possibility could have developed if freshwater fishes, which gave rise to amphibians and all subsequent vertebrate life, had not started trundling across those shallow ponds that covered the shores of the supercontinent Pangaea 400 million years ago. A mere absence of such ponds probably would have been sufficient to prevent the conquest of the land by amphibian descendants of those fishes—stressing once again the narrowness of the crack in the evolutionary door through which all mammals, including man, leaped onto life's stage.

A world particularly suitable for insects would be a steamy, jungly world. "Insects own the jungle," says the noted Harvard biologist Carroll M. Williams. "And if you want to see insects on the most massive scale, it would be in a tropical rain forest."

While it is true that insects on Earth have adapted to many habitats—from the desert sands to the faces of the glaciers—most insects are found in the tropics. There are good reasons. Insects need warmth to survive, and in the tropics they are less specialized and therefore have a better chance of changing into a different form than do their more specialized relatives living elsewhere.

If insects had had the opportunity to fill an evolutionary gap—to fill the niches left open by the absence of amphibians, and later reptiles and mammals—they could have easily poured into the openings. As we have seen in Chapter Four, "The Twists and Turns of Evolution on Earth," a single pouched opossum gave rise to Australia's diverse animal world. Similarly, insects could have radiated into niches now filled by other animals.

If the Earth had a chance to become The Planet of the Insects, such a situation more likely than not would have arisen on innumerable alien worlds.

Insects can readily be visualized as soaring hunters on the plains, grass eaters, carnivores, and other counterparts of birds and mammals —all greatly magnified in comparison with the Earthly models. They would create a world filled with chirping, clanking of mandibles, the whistling of bird-like insects swooping toward the ground—perhaps a nightmare world but a real one nonetheless.

Another type of insect-dominated world can be visualized where mammals and other vertebrates are held in check by the overwhelming preponderance of insects. In that world, the roles of insects and vertebrates would be reversed: huge insects could hunt tiny mammals.

There are insects on Earth that prey on higher animals. The larva of the dragonfly, for instance, matures underwater, where it attacks and devours fish and tadpoles twice its size. Army ants will fight and try to overpower any kind of life that gets in the way of their single-minded march. A big spider that lives in Australia weaves nets strong enough to catch small birds.

On Earth, it's only their relatively small size that has prevented insects from literally taking over the world. If that takeover had taken place, insects would have been certain to move into the niches now occupied by birds and mammals. The humming moth, which looks and acts like a smaller hummingbird, provides an example of such expansion even in competition with birds.

A world dominated by insects would include large, intelligent creatures like those shown in our color insert. To be sure, for insects to grow much bigger than the terrestrial models a restructuring of the insect body would be in order.

Two factors keep the size of insects down on Earth: Their skeletons are on the outside as if the insects were sitting inside armored cars, and because they have no lungs, they must pump oxygen to their bodily tissues through air tubes. Doubling an insect's size under these conditions would cube the mass of its tissues, making them too heavy for the chitinous exoskeleton to support and require more oxygen than the air tubes could supply.

Yet there is no reason why an insect-like creature on another planet could not evolve a large body that works, as well as a good brain. (Note what a bee can do although its brain is no bigger than a lentil seed.)

If, instead of air tubes, insects used their heart-blood circulation, which they have, for respiration, they could evolve into bigger and better creatures and take the path to higher intelligence. With reptiles, birds, and mammals present on Earth, insects couldn't do that. They are repressed and have taken the road to vast numbers of small creatures, much the same way, for instance, as female octopuses, which lay hundreds of thousands of eggs at a time to ensure that a few offspring survive the predators of the seas. Without such repression, the way would be open.

Under certain circumstances, insects could grow much bigger even without changing their method of respiration. On a planet with a dense oxygen atmosphere, for instance, insects could supply oxygen to their tissues in greater volume and could thus grow to much larger size. On such a planet, a premium would be placed on taking to the air. Accordingly, the air may be filled with flying creatures of every description—but shaped and functioning much like airplanes in keeping with the dictates of the environment.

The other way for insects to grow bigger would be to take the route their crustacean relatives did. Both crustaceans such as lobsters and crabs, and insects, are arthropods, members of a phylum that also includes spiders, ticks, millipedes, and centipedes. With more than nine hundred thousand species, and new ones being discovered all the time, insects are the largest group of arthropods and account for nearly half of all the animal species on Earth.

Arthropods are animals with jointed legs, which is the meaning of their name in Greek. Both crustaceans and arachnids, such as spiders and mites, differ from insects in that they have eight legs, compared to the insects' six. In addition, the heads and chests (thoraxes) of crustaceans and arachnids are fused into a cephalothorax and abdomen; insects' bodies are divided into head, thorax, and abdomen. And insects often have wings and a pair of antennae. The biggest difference between insects and crustaceans, though, lies in the fact that crustaceans got tooled up with circulating blood, and evolved gills instead of the insects' tracheal tubes. This has enabled lobsters, for instance, to grow to forty pounds and more, a size Earthly insects never attain. (The crustaceans' blood pigment, incidentally, as indicated earlier in this chapter, is hemocyanin, which is built around the copper atom, rather than the iron atom, which colors our blood pigment, hemoglobin, red, in contrast to the bluish tint of hemocyanin). The lobster grows and molts, or sheds its shell, and has a life-style not unlike aquatic insects, except that it has a better respiratory system.

The question may be raised whether an insect evolving to a more complex state, or at least to a large frame, would still remain an insect or become a "flesh and blood" creature like a crab or a lobster. That would probably depend on the circumstances on the particular planet. With high oxygen content in the atmosphere, insects could retain their familiar features. With lower oxygen content, on the other hand, they could take on the crustacean countenance, although not necessarily remaining crawling creatures. They could rise on their hind feet to walk the land, evolving into somewhat grotesque counterparts of man and other mammals.

Whatever their shape, insects could advance much farther in intelligence. Mistakenly, insects are often labeled a lower form of life. They are far from being that. The complexity of their anatomy, the performance of their organs, their unique senses, their remarkable societies and cleverly constructed abodes, match and even exceed those of higher vertebrates.

On Earth, insects have been the most successful of all living things.

Just how far insects could advance is hinted by some of their remarkable activities on Earth.

Insect societies, for example, display such sophisticated attributes as the division of labor and the social structure in the beehive or the anthill, as well as the striking spectacle of long columns of army ants on their foraging marches.

Other ants act as farmers by keeping aphids as "cows." They caress the tiny insects to make them release honeydew, on which the ants feed. What's more, these types of ants even build protective shelters of paper-like material around the plant stalks on which the aphids feed; they move the aphids from plant to plant, and even protect aphid eggs during the cold months by taking the eggs into their underground nests. Although based on instinct, this type of activity has the makings of intelligent operations.

In the social structure found inside the palaces of bees, wasps, ants, and termites lie the makings of even more complex societies, which could be elaborated on by intelligent insects elsewhere.

Crustaceans, being arthropods, can be viewed as superior insects. They suggest an avenue to higher intelligence that insects may take elsewhere. Like insects, crustaceans display reflex behavior such as opening and closing of claws, defense, escape, and feeding. That much one might expect. But it may come as something of a surprise that crustaceans also engage in elaborate courtship rites and practice advanced parental care. Fiddler crabs, for instance, practice organized feeding and have dominant males in family groups and ceremonial fighting among males. Many crustacean species communicate by means of visual signaling and by sounds produced by rubbing their legs together.

To talk about insects' mental capacities may seem strange, yet scientists have successfully trained bees to walk through complex mazes, responding to various cues such as the color of a marker, and making as many as ten turns in sequence in response to such cues.

Furthermore, scientists have determined that bees have good memories. They can remember the location of a food site for six to eight days, and perhaps for as long as two to three months.

Ants can be taught to run mazes too. And they learn only two or three times slower than rats.

In fact, it has now· been shown that social insects can be taught many of the same things that higher animals can. Insects respond to habituation, for instance—they get accustomed to stimuli. They are capable of associative learning—they can associate reward with stimuli that earlier were meaningless to them.

Such feats by insects may sound surprising, even somewhat incredible, until one realizes that while the insect brain is small, insects make up for its smallness with structural compactness. The neurons in the insect brain are smaller in size than those of vertebrates, so more of them can be packed into the same space. Connections between neurons also are more compact.

Insects, furthermore, compensate for the small size of their central brain by spreading their nerves and sense organs around their bodies. Insects, for instance, can tell temperature with their feet, hear with special organs on their legs, and detect scents from miles away with their antennae.

Tiny hairs on the insects' bodies perceive vibrations in the air and ground to warn of approaching danger. Thanks to this distribution of the nervous system and sense organs, each of an insect's three body parts—the head, the thorax, and the abdomen—are semi-independent. They can carry out reflex actions on their own, without communicating with the brain.

To be sure, although this compensation hints at the possible distribution of sensory organs in creatures on other worlds, it has not been quite adequate to bring insects on Earth to the mental level of even the simplest vertebrates. The insects' brains are still too small, although they have obviously served them well enough to make them the most successful animals on Earth. But while insects probably match vertebrates in innate behavior, their capacity for learned behavior is vastly inferior. When confronted with novel problems, insects are generally incapable of reorganizing their memories to deal with new situations. Insects taught to run mazes, for instance, can't run the same maze in reverse, but have to master it as a new problem. Rats and other vertebrates, on the other hand, make use of such previously acquired information.

This suggests that social insects inherit whole blocks of behavioral patterns. They are not quite the "reflex machines" scientists once thought them to be, however. Within a certain range of possibilities, insects have a prescribed choice of reactions that need not always be exactly the same but that nevertheless lead to the same results.

Despite their varied ways of life, there is not much change or flexibility in the workings of insect societies as they have evolved on Earth. In a number of respects, to be sure, insect societies outdo those of man, particularly in terms of cohesion, specialization of the members of the castes, and in individuals' single-minded dedication to the common good. Still, while the structure of insect societies effectively insures the survival of the species, it doesn't lead to any advanced forms of social interaction. Insect societies on Earth seem to be frozen in development. Possibly, insects here have already exploited all the social strategies within the limitations of their brains.

This could be different on insect worlds elsewhere, of course. Larger brains would give insects there greater flexibility to build new social relationships on the existing instinctual foundation. For one thing, the individual's role could become much more important, and relationships between individuals could assume more-human forms. One example, that of an insect mother handing her daughter a flower, is shown in our illustration.

To say that is not simply to attempt to ascribe human-like activities and motivations to other life-forms. We already see some interesting parallels to human behavior in the activities of insects, particularly in their use of tools and in their construction work.

But there is more. Insect altruism, the readiness of a soldier ant or a worker bee to sacrifice its life for the good of the colony or hive, is another example of human-like behavior.

This kind of similarity points to interrelatedness of all life. We don't have to make the universe anthropomorphic—it is anthropo-

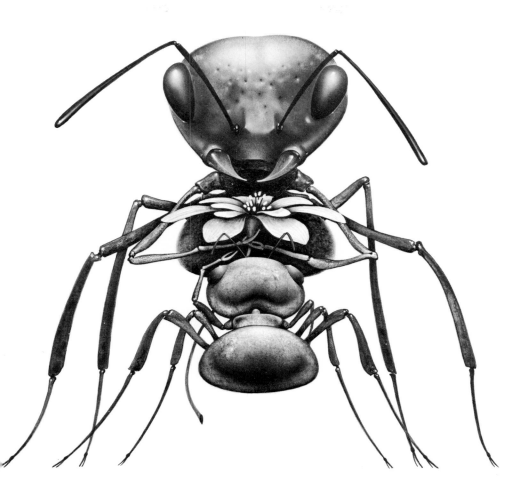

morphic, at least in some respects. Behavior is rooted in biology, as the new science of sociobiology is demonstrating, and a lot of this behavior takes similar expressions in all animals, ranging from ant to man.

In that light, consider the Earthly insects' construction abilities. In terms of types and varieties of structures, the world of insects holds fascinating possibilities that insect-like creatures elsewhere could enlarge and improve upon. Examples of clever structures built by insects on Earth abound—structures not only to live in but to trap prey, too.

Ant lions, larvae of an insect that looks like a dragonfly in its adult stage, for instance, build funnel-shaped pit traps in sand or dust to catch ants.

Once an ant is captured, the ant lion then engages in a unique dining operation. It first injects a deadly poison into its victim. Then, instead of dismembering its prey, a task for which the ant lion is not equipped, it injects its gastric juices into the victim. These literally dissolve the soft parts of the ant. The ant lion then elegantly ingests the dissolved nutrients through a groove in its mandible. Since its food is digested outside its body, there are no residual substances to dispose of—an elegant way of feeding that may be employed by much larger insects on other planets.

Some ants use extremely imaginative tools to construct their dwellings. Weaver ants in tropical Asia build nests in trees by tying together undetached leaves with dense silky webs. Since the ants have no spinning glands, scientists wondered for a long time just how weaver ants built their structures. They finally discovered that the ants first pull the edges of the leaves together to form a living bridge of ants to cross the gap too wide for a single ant to cover.

Then, in what must be one of the most remarkable demonstrations of tool-using in the animal world, worker ants bring full-grown larvae from deep in the nest. Holding the larvae in their mandibles much as a human would hold a plastic tube of glue in his hands, the ants press the mouths of the larvae against the leaf surface, at the same time forcing the larvae to discharge silken threads to stitch together the edges of the leaves being held in place by other worker ants.

If worker ants on Earth can use tools in such an intelligent manner, it isn't too much to expect much larger, more intelligent insects elsewhere to use tools as well as man does.

Similarly, intelligent insect-like creatures on other worlds could far outdo the master builders and architects among insects on Earth. Termites are such master builders. Their nests can reach giant size, towering like strange obelisks over the landscape. The compass termites of Australia, for instance, build impressive structures up to fifteen feet tall and nine feet wide in the treeless prairie. In human terms, these structures would tower to nearly a mile, four times as tall as the Empire State Building.

The structures' short sides face north and south, making the surface exposed to the hot midday sun small. The long sides are exposed to the cooler morning and evening sun. Like bees, termites sense magnetic forces to orient their skyscraper dwellings.

In the tropical rain forests, on the other hand, certain termites build umbrella-like roofs over their dwellings to keep the rain out of the main structure.

The air-conditioning systems of termite dwellings are marvels of engineering. Since some of the mounds are populated by millions of oxygen-breathing termites, without ventilation the termites would soon suffocate. So the termites build elaborate systems of air ducts and spaces to allow an exchange of carbon dioxide for oxygen.

Enlarge the activities, structures, and societies of Earthly insects, and a fairy-tale world takes form. We can almost hear the clanking of the mandibles as insects resembling medieval knights in their armor engage in combat for survival. Monstrous creatures fill the skies; bird-like insects that prey on slower, ground-based compatriots, not only swooping down on them like hawks but also shooting arrow-like stings.

Without predators of other types, insects will fight insects. The head appendages of insects around the mouth are made of hard material—mineralized chitin. Insects form this material into all kinds of fantastic mouth parts to become highly specialized in both feeding and fighting. Soldier ants offer particularly striking examples. Some have evolved mandibles shaped like curved daggers with needlepoint ends. Snapping them together causes a bite painful enough to deter even larger vertebrates from bothering the ants. Other soldiers are equipped with glands that manufacture sticky or poisonous substances that they can squirt at enemies inches away. The possibilities for even more clever weapons are there.

Insect-like creatures elsewhere could also escalate the already highly refined forms of parasitism and sneak attack that exist among insects on Earth. Some insect species use others as mobile food stores for their offspring by depositing their larvae into other insects. Where sneak penetration is concerned, few animals can match the perfidy of certain ant queens. By inciting foreign worker ants to carry them into the foreigners' nests, by feeding the ants tasty substances the queens manufacture, the intruder queens take over the colonies by strangling the reigning queens or by literally cutting off their heads.

Little wonder that insects seem otherworldly. "Something in the insect," wrote Maurice Maeterlink in *The Life of the Ant*, "seems to be alien to the habits, morals and philosophy of this world, as if it has come from some other planet, more monstrous, more energetic, more insensate, more atrocious, more infernal than our own."

Already equipped with sensory capabilities that we don't have—such as the ability to orient their flight by polarized sunlight, to measure the thinness of their cell walls to a fraction of a millimeter, to see in ultraviolet light, among other things, insects elsewhere could construct cities and other structures of immense size and great beauty.

They could make and use tools and perhaps even develop technological civilizations.

The mastery of fire, a prerequisite of technological development, might be achieved by insects elsewhere in a way more elegant than primitive man's. Instead of rubbing together two sticks to produce enough heat to start a fire, an intelligent insect might do it by capturing the sun's rays with lenses made of transparent gelatinous substances produced by the insects themselves. With appendages skilled in ways that our hands aren't, these intelligent insects could build tools and instruments that took us centuries to produce because of lack of appropriate machine tools.

Size aside, insects can proceed to grow large brains, develop better respiratory organs, and significantly extend their life-spans. Building upon their unique senses, insects could emerge as a kind of biological machine of a high order, the kind of advanced robot science fiction writers love. These otherworldly insects could be as big as humans, and they could see in ultraviolet, detect sound through their feet, keep precise time, sense magnetic fields—in short, they could be born scientists.

Weaving flight tunnels through the treetops to protect themselves from both predator insects and the elements, such creatures could construct elaborate systems of interconnected habitats where they could engage in complex activities. (See color insert.)

Strange collaborations between differing insects could arise. On Earth, the great armies of soldier ants, which are usually blind, are accompanied by a motley crew of odd camp followers like the millipedes that follow army ants, marching in the middle of the column; beetles that strikingly resemble their hosts in shape and form; and still other beetles that seek adoption by ants by presenting workers with "appeasement" and "adoption" substances from their glands and, as a result, are carried inside the nests, where they live.

We could also visualize greatly enhanced agricultural activities by insects. Ants of the genus *Atta* provide an example. They culture and eat tiny fungi, relatives of the mushroom. Many species of this ant cut pieces of fresh leaves, which they then carry into their subterranean dwellings. Other small worker ants of the same species accompany the leaf-cutting ants; their sole function is to repel the attacks of parasitic flies, which lay eggs in the necks of the leaf cutters. The protective workers ride home on the pieces of cut leaves being carried like parasols by the leaf cutters, awaiting attacking flies with their jaws open.

Once home, the small workers begin to tend the fungus gardens, one of the most intriguing inventions in the insect world. The workers chew the leaf fragments into small pieces, mix them with saliva, and fertilize them with their feces. In this spongy material, the fungus is then cultivated by the worker ants. The workers even bite off branches of fungi so that the plants will develop cabbage-shaped heads, the delicacy on which the ant colony lives. That particular strain of fungus is passed on from ant generation to ant generation. When a queen founds a nest, she brings with her in a pocket of her mouth cavity a small piece of the fungus.

Such complex activity obviously has the makings of agriculture, which could well be expanded and built upon by insects elsewhere. It shows once again that evolutionary convergence extends not only to animal forms but also to animal activities. Ants and men are as distantly related as any creatures on Earth. Yet here they are, both engaged in agriculture.

Any insect world would be rich in plant life, because insects depend on plants for food and many plants depend on insects for pollination.

Some plants, such as conifers and wheat, for instance, rely on the wind for pollination, to be sure. They have inconspicuous flowers. But plants that depend on insects usually display brightly colored flowers and often secrete nectars. The evolution of insects evoked the world of brightly flowered plants.

The interrelatedness of plants and insects in alien environments would be one of the most fascinating aspects of an insect world.

Consider such relationships on Earth. Some Earthly plants live in a strange symbiosis with insects. Certain tropical plants, for instance, provide specially designed dwellings for ant colonies—chambers inside hollow stems stocked with fats and proteins to attract the ants. The ants serve as the plants' standing armies, repelling damaging insects and even cattle and other mammals, which they sting.

Some insects, in turn, take advantage of plants by depositing their larvae on leaves or branches. The larvae secrete growth substances that make plant cells grow around them in protective galls sometimes as big as baseballs. As it grows, the larva either pushes, or eats, its way out. In at least one case, the larva's growth substance programs construction of a plug in the gall that pops out when the larva is fully grown.

Not all the relationships are peaceful or mutually beneficial. Some plants on Earth have turned the tables on insects by becoming insecteaters. There are about five hundred kinds of carnivorous plants that capture insects by shutting their leafy traps, by immobilizing their prey on trap surfaces as sticky as flypaper, and by drowning insects in pitcher-like flowers filled with water.

Another technique to capture animals, so clever that it looks as if a higher intelligence designed it, is employed by microscopic fungi. They are equipped with tiny handcuff-like snares that shut tight when a microscopic worm known as a nematode crawls through one. Trapped by as many as three such snares, the nematode is firmly held in place for fungi filaments to invade and kill it.

Now imagine an alien world where such relationships are vastly magnified. A battle of wits between the plants and the insect legions could result, to outdo any such encounters on Earth.

A battle of wits between plants and insects?

But why not?

We shouldn't define intelligence too narrowly. Intelligent plants most likely will not only be recognizable as plants but will also be equipped with sensory systems that would make them *appear* intelligent. There are plants on Earth that react to touch, heat, and other stimuli nearly as much as some insects do. "A praying mantis sits there immobile most of the time," says Peter Carlson, a pioneering plant geneticist at Michigan State University. "And then, suddenly, it reaches out for a chunk of passing food. It's an instinctual reaction and I'm not sure how different that is from a plant turning toward light."

Furthermore, there are plants in the sea that act like animals.

But how could plants develop nervous systems? Listen to Peter Carlson again: "I don't think that we should define a nervous system only as we see it in higher animals. We see a spinal column, nerve endings, and specialized cells. But plants also have ways of communicating. Their cells talk to one another by using a language called hormones. You may think of that as a very slow process, and it may be, relative to the speed of an electric impulse of a nerve cell. But I can imagine a more rapid form of communication with hormones and, as far as I'm concerned, that would constitute a nervous system. You don't have to have nerve cells and brains to have a nervous system. All you need is a way to pull it together and make the message interpretable and have the plant react to it."

Equipped with such nervous systems, plants on other worlds could be much more active than plants are on Earth. On The Planet of the Insects, they could interact with insects much more aggressively than they do on Earth, both in collaborative and in adversary relationships.

Plants could also populate a planet without any insects or animals being present. A number of plants on Earth are either self-pollinating or rely on the wind for pollination. So plants could perpetuate themselves without help from animals, including insects. And where the photosynthetic cycle is concerned, plants don't need help from animals either, because plants themselves complete the cycle: they produce both oxygen and carbon dioxide, which they recycle.

Without animals to consume them, plants could expand into all the vacant niches normally filled by animals, with some plants even acquiring a degree of locomotion and others preying on other plants. We know through our own manipulation of plants by breeding that the genetic potentialities are there and that the plants on Earth are continuing to evolve. So, on The Planet of the Insects, plants may well hold their own against insect hordes and even surprise them by evolving into unexpected forms.

The intelligent octopus, intelligent insects, even intelligent plants are only a few of the multitude of intelligent creatures likely to be found in the Darwinian universe. There are bound to be planets younger than the Earth, to be sure, where life has not gotten beyond one-celled bacteria. Some older planets may have only such life even now; this could have happened on Earth if our planet had never developed oxygen-breathing, one-celled plants that helped create the atmospheric conditions needed for higher life as we know it to evolve.

Those primitive planets obviously would be less interesting than the ones harboring intelligent organisms based on building blocks familiar to us. But could those creatures follow life plans entirely different from ours? Is there such a thing as "life as we don't know it"? That's the subject of our next chapter.

CHAPTER SEVEN

Life Beyond Darwin

The universe, according to many science fiction writers and some scientists, is populated by a bewildering variety of grotesque, bizarre, and monstrous creatures that have little, if anything, in common with our kind of life.

The theme that has been followed by many contemporary science fiction writers was sounded by H. G. Wells at the end of the nineteenth century in his classic *The War of the Worlds* when he described the invading Martians in these words:

> They were, I now saw, the most unearthly creatures it is possible to conceive. They were huge round bodies—or rather heads —about four feet in diameter, each body having in front of it a face. This face had no nostrils—indeed the Martians do not seem to have any sense of smell—but it had a pair of very large, dark-colored eyes, and just beneath this a kind of fleshy beak. In the back of this head or body—I scarcely know how to speak of it— was the single tight tympanic surface, since known to be anatomically an ear, though it must have been almost useless in our denser air. In a group round the mouth were sixteen slender, almost whiplike tentacles, arranged in two bunches of eight each.

Wells's Martians also displayed a predilection for human blood.

Passage of time has not improved the visions of cosmic life-forms. Having names made up of jumbled consonants that may sound like music in the bazaars of Baluchistan but grate on our ears, the current inhabitants of science fictional realms strain our credulity.

Take Chulpex, for instance. It is a four-armed entity that lives in a labyrinth of tunnels deep in the interior of its planet.

Or take Czill, a mobile intelligent plant that can speak.

Or Dirdir, a monsterish hoofed creature with antennae extending from its head.

Or Ixtl, which has a cylindrical body with four legs and four arms and emanates a mysterious force that allows it to pass through solid matter.

Or Pnume, a skeleton-like creation that can walk on two feet or become quadruped when it so desires.

Or Thrint, a one-eyed, telepathic, space-suited frog-like monster with needle-sharp teeth that turns into a Ptavv if it doesn't pass muster.

Or Ruml, a furred cat-man that likes to duel.

Or Uchjinian, a noncarbon-based pliable glob of matter that lives on a planet with a helium atmosphere.

That's just the beginning of the parade.

Blithely disregarding the basics of biology as well as science in general, some science fiction writers present us with rock-like, sedentary creatures that somehow communicate via radio (but only after gaining permission to transmit from a "queen," who sounds like a one-woman Federal Communications Commission); and with combined plant-animals that look like puppets from Sesame Street; with three-legged creatures with webbed feet and hands, who despite those handicaps still somehow manage to build spaceships that fly at 99 percent of the speed of light. And even at this late date, when silicon monsters have been laughed out of the cosmic court, some science fiction writers still try to palm off on us silicon creatures such as intelligent trees made of glass.

But if science fiction writers have fostered the belief that creatures on other planets will be unbelievably grotesque, Hollywood has embellished them with mindless aspects of horror, or, at the other extreme, portrays them as mirror images of man.

The ironic fact is that the monsters and oddities that the speculators see in their cloudy minds' eyes are nothing compared to what nature has supplied on Earth, probably unbeknown to the fantasy fraternity and to the movie producers.

We show some of these surprising products of evolution, some of them still living, in the illustrations that appear throughout this chapter:

These drawings appear to justify the remark by Dr. Berrill that "Everything you could possibly imagine, you'll find that nature has been there before you."

In fairness to science fiction writers, to be sure, there are some no-

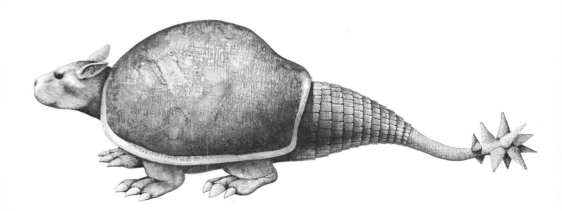

TOP: *The armadillo-like* Doedicurus, *which once lived in South America.*

BOTTOM: *The rhino-like* Styracosaurus, *which once roamed North America.*

table exceptions to the general paucity of good ideas about life-forms elsewhere in the run-of-the-ray-gun romances. Arthur C. Clarke comes through with some believable biological robots in his haunting *Rendezvous with Rama* (Ballantine Books, 1973). Some lesser-known writers also present plausible life-forms.

Ursula K. Le Guin's Athshteans, in her *The Word for World Is Forest*, for instance, are furred, man-like beings that evolved from tree-dwelling primates.

Gordon R. Dickson's Dilbians, in *Spacepaw*, are bear-like intelligent creatures.

James Blish's Lithians, in *A Case of Conscience*, are bipedal reptilian entities whose life cycle recapitulates their evolution from fish through amphibian to a kangaroo-like lower mammal to mature adult.

In Brian M. Stableford's clever *Wildblood's Empire*, the amphibious entities he calls the Salamen have the choice of changing either into an adult aquatic form or juvenile air-breathing land dwellers.

But those intriguing elaborations of biological possibilities are the exceptions to the general run of cosmic nonsense. The reason science fiction writers for the most part fail to come up with believable beings is that their speculation, more often than not, is not based on an expansion of evolutionary possibilities. It is based, instead, simply on whatever odd or grotesque idea happened to drift into the author's head at that moment. Plots are solved as the need arises by devices that disregard scientific possibilities and limitations. There is obviously no science in that kind of fiction.

But science fiction writers are not alone in their wild imaginings. They share with some scientists the dubious distinction of creating cosmic myths—probably indicating that at least some scientists are just as susceptible to unsubstantiated speculation as are laymen. Such speculation has a long history, but we'll mention only the perhaps most intriguing examples.

Incredible as it may sound, as late as the nineteenth century, for instance, some of the greatest scientists then living entertained the idea of life on the sun. Theories were devised to explain the sun's structure on the basis of the existence of solarians, mythical inhabitants of the solar orb.

Earlier scientists had been particularly puzzled by the nature of sunspots. The spots seen on the sun usually consist of dark portions called umbrae. Around them are gray rims known as penumbrae. At times, though, the black spots appear without the grayish border, or

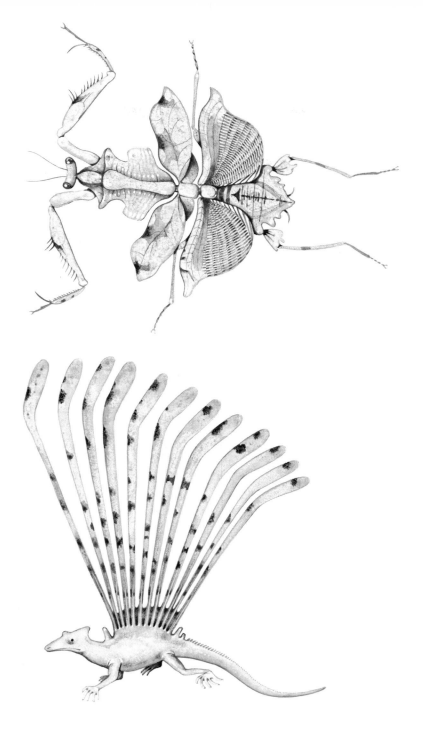

TOP: Deroplatus sarawaca *is an insect that looks like a Rorschach test and lives in Borneo.*

BOTTOM: *The long-extinct* Pseudosuchians longisquama *could easily pass for an extraterrestrial creature.*

the spots themselves look gray. To complicate the picture, bright, cloud-like patches known as flocculi are often associated with the spots, although they also occur independently.

As usual, scientists sought to explain the mysteries. The attempt was begun in 1774 by Alexander Wilson, professor of astronomy at the University of Glasgow, and carried to conclusion by Sir William Herschel, the illustrious founder of sidereal astronomy and discoverer of the planet Uranus.

A description of Herschel's imaginative explanation is interesting not only for its own sake but also because it provides a historical link to similar speculations today on the nature of life different from ours, also by seemingly serious scientists.

Herschel saw the sun as a dark sphere swathed by two distinct atmospheric layers. A different kind of clouds floated in each layer. In the lower envelope, the clouds were dense and opaque, much like the clouds in the Earth's atmosphere, reflecting light. Above them floated phosphoric clouds, which gave the sun its luminosity. Herschel likened these clouds to the northern lights in our atmosphere but extending much farther into space.

The solar atmosphere and its natural changes thus offered an explanation for the sunspots. It all fell neatly into place, according to Herschel. When observers on Earth saw the dark spots, they were seeing through the two layers of clouds all the way to the solar surface. When they saw the grayish spots, on the other hand, they really saw the top of the opaque bottom cloud layer in places free of the luminous upper clouds. Finally, when they saw the flocculi, what they really saw were greater concentrations of luminous clouds.

That wasn't all. Below this fanciful but, for its day, cleverly constructed atmospheric pyramid, Herschel populated the sun with a happy race of beings who lived a trouble-free life in a climate of eternal spring. The opaque cloud layer protected them from the heat and light of the luminous clouds high above, giving them at the same time perpetual daylight and a climate that didn't vary from one pole to the other. Through the openings in the clouds—sunspots to us—the sun's inhabitants could glimpse the universe. "The sun," concluded a follower of Herschel's, "may easily be conceived to be by far the most blissful habitation of the whole system." If nothing else, this was a highly imaginative and rather charming picture.

Similarly entertaining mythology evolved about terrestrial planets of the solar system, so called because until recently they were believed to be much like the Earth. As recently as thirty years ago,

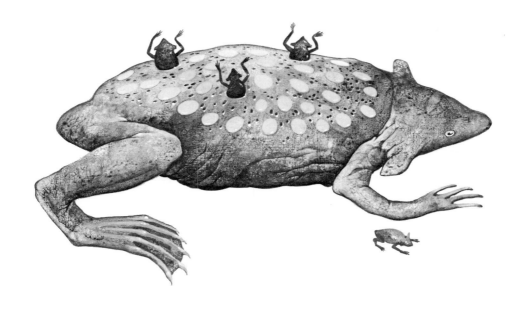

TOP: *The Surinam toad has devised an unusual way of carrying its young.*
BOTTOM: *The deep-sea angler fish is ferocity personified.*

some astronomers were expounding the idea that planets were formed in reverse order, starting with the outermost, Pluto, and proceeding successively to the innermost, Mercury.

In this scheme, Venus was placed at a stage of development millions of years behind that of Earth, and Mars became an ancient planet.

In keeping with that idea, the planets were populated with beings that would suit the planets' ages. Since the inhabitants of Venus supposedly were younger, they were given frivolous traits. In 1951, for instance, Kenneth Heuer, an American astronomer, in discussing the riddle of the ashy light on Venus, a weak glint that shines in the nonilluminated regions on the night side of the planet, reached a remarkable conclusion. Heuer stated that until the riddle is solved, he would "much rather believe with Gruithuisen, a nineteenth-century director of the Munich Observatory, that the faint light of Venus is produced by festivals and general illuminations organized by the inhabitants of the planet at certain epochs. These festivals are probably celebrated on the occasion of political changes or according to religious periods."

In a similar vein, Heuer wrote about Mars: "Perhaps there is a vast subterranean civilization connected by an elaborate system of tunnels."

So as not to leave the asteroids uninhabited, Heuer placed on them "living minerals—singing stones and talking rocks. . . ." (*Men of Other Planets,* by Kenneth Heuer, Collier Books, 1963.)

Heuer saw Jupiter as a world populated by weirdly fanciful animals and plants.

You are wrong if you think that we have progressed very far from Herschel's or Gruithuisen's days in the quality of scientific speculation about nonbiological life that may exist amid the stars.

The current speculation, to be sure, is dressed up in new scientific garb, but its end result is often just as amusing as Herschel's solarians.

We must look briefly at our own kind of life again, though, to see what the new speculators are leading up to.

Life as we know it, by definition, is life based on carbon and water.

The requirements for a biological system are so stringent that only certain types of elemental substances could ever fit into the category as the necessary building blocks for that structure. In order for an organism to survive, it has to have a minimum number of catalytic structures within its environment. If it doesn't, then it must compete with its environment. It must concentrate the environment, and that takes energy. In order to have that type of energy, you have to have a multiple set of complex substances in order to do the job. So we simply can't envision any structure outside of the carbon chemistry that could handle the job.

As we saw in Chapter Four, "The Turns and Twists of Evolution on Earth," both silicon and boron, the two often-suggested substitutes for carbon, are limited in their abilities to enter into the necessary number of complex formations. Carbon, on the other hand, is unique. It is one of the most abundant elements in the universe, as are the three other major constituents of life: oxygen, hydrogen, and nitrogen.

These are also elemental substances that tend to seek the surface of any planet. Being light, they tend to move up from the core—an obvious reason why nature chose these elements. Once a living system is organized, energy must be pumped into it. The only way that solar energy can get to any planet is through the atmosphere—or at least the appropriate kind of solar energy, with its harmful components filtered out. So the activities of life have to occur on the surface exposed to some energy bank, some source of energy, and that must be radiant energy. Thermal energy is not quite sufficient to crank up a living system. You can't just get a system started by pumping heat into it. You have to pump heat out of it in order to get energy into it, and that's a very difficult job. As a result, it is only under very special circumstances that we have life on anything. That special circumstance is, of course, met by our environmental conditions on Earth.

How can such conditions be met on the surface of a star or on an extremely hot planetary surface? In fact, can they be met at all to produce what some scientists inelegantly call "life as we don't know it"?

Some specialists have manfully struggled with this difficult idea. V. A. Firsoff, a British astronomer, has tried to visualize worlds where creatures breathe fluorine and drink liquid hydrofluoric acid and where other beings have substituted hydrogen sulfide for water. On a planet with high atmospheric pressure and high temperatures—reaching thousands of degrees Fahrenheit—Firsoff sees creatures made of silicon.

But Firsoff has been forced to conclude, as he concedes in his book

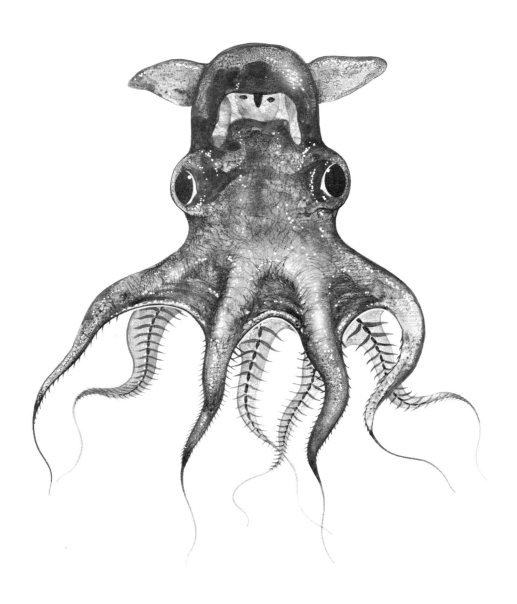

The vampire squid fortunately does not grow larger than about fourteen inches and lives in deep waters.

Life Beyond the Earth, which appeared in the 1950s, that "not all such solvents, even if theoretically satisfactory, are likely to exist in sufficient profusion on a planetary surface at the right temperature to become the basis of a life system." He concedes that many of the conditions he paints are only a remote theoretical possibility. Fluorine, for instance, is so reactive that it would soon become chemically bound and disappear from a planet's surface.

Since fluorine is too reactive, chlorine has been suggested as a substitute for it. In this scheme, a planet would contain tetrachloride in its atmosphere, which plants would utilize as a counterpart of carbon dioxide in the terrestrial air to make a carbon-chloride counterpart of starch.

While alternative, or at least somewhat different, chemistries can't be completely excluded, the prevalence of carbon and water in interstellar space speaks against such chemistries. When nature has a choice between elegant and inelegant solutions, it invariably chooses the elegant approach. In this scheme, leathery creatures with fire-resistant skins and gullets that withstand acidic liquids are inelegant solutions, while carbon-and-water creatures are elegant.

Furthermore, as we have seen, it's very hard to duplicate the unique aspects of carbon-and-water life with alternative chemistries. Not even molecules that replicate can be constructed of such substitutes, much less any organisms. It's hard to see, therefore, how such alternative chemistries can lead to life of any kind.

Such considerations, however, haven't stopped some scientists from suggesting even stranger life-forms. Aside from alien chemistry, some physical scientists today are making strenuous attempts to show how life with a physical, rather than a biological, basis may have evolved elsewhere—a modern variation on Herschel's solarians.

Life, of course, as well as being chemical *is* physical in the sense that all processes of life familiar to us are based on electromagnetism. "It is the electromagnetic field that holds the harem of electrons about each nucleus to form atomic systems," says the noted Yale physicist D. Allan Bromley. (The electromagnetic forces arise from the positive and negative electric charges that the atoms contain). "It is the electromagnetic field that holds atoms together to form molecules, including the marvelously complicated molecules of life itself, and it is electrochemistry that activates the structure of any living system of significant complexity."

But the scientists who propose life based on physics rather than biology consider an expanded system in which the contact between individual molecules, which characterizes the living systems we know, is replaced by a much more diffuse arrangement. In this idea, individual atoms and molecules that might be considered to be participants in the organism are linked over long distances by electromagnetic fields. Notes Dr. Bromley: "This, of course, immediately poses a fundamental problem, depending upon the scale envisaged, because inasmuch as no information can be propagated at velocities exceeding those of light we either have an upper limit on the size of the organism that can be considered or a very sluggish beast at best."

Among the recent suggestions are beings composed of energy instead of chemicals—certainly not a new idea in science fiction. But the idea tends to acquire some respectability among laymen when pushed by scientists, even though that is no guarantee, of course, of such systems' plausibility.

Herschel and his solarians may be dead, but the idea of life on stars, and even inside them, isn't. This modernized version ostensibly is based on well-established principles of physics and thermodynamics, the branch of physics that deals with the relationship between heat and work. (In the color insert, we show a creature that lives by withdrawing energy from a plant in a novel way.) Heat, of course, is a form of energy that results from the motion of molecules in a substance.

The first and second laws of thermodynamics are frequently invoked by scientists who speculate about physical life (or generalized life, as some of them have called it). The first law states that energy can neither be created nor destroyed, only changed from one form to another. Thus, a definite amount of mechanical energy can be changed into heat, and vice versa.

The second law essentially states that the energy engine of the universe is running down, in view of the fact that heat cannot of itself pass from a colder to a hotter body.

Within that larger universal setting, however, this running down can be stopped in complex systems if the available energy is positive. Such a situation prevails on Earth, where the sun serves as the source of energy for the growth of vegetation and subsequently for the maintenance of animals that consume that vegetation. The environment, on the other hand, serves as the heat "sink" for both plants and animals to discharge waste heat.

Thermodynamics not only helps explain how such plant and animal heat engines come about but also suggests a rationale for physical "heat machines" resembling living things, argue these scientists. Once a system starts storing energy, the system becomes highly organized. If such a system, furthermore, begins to exercise control over some property and begins to manipulate it, such a change will be a manifestation of behavior. Such cumulative effects can then lead to what amounts to information storage, and from there to rudimentary aspects of a life-like organism.

The proponents of physical life, or more properly of patterns and configurations that may have life-like qualities, argue furthermore that storage of information in discrete chemical units as it takes place in our type of life is only one way of doing it. They argue that thermodynamic principles allow the evolution of organized information systems based on physics, rather than chemistry.

Such systems, goes the argument, can exist in a variety of surroundings—from the unimaginable heat of the interior of neutron stars to the frozen surface of Jupiter. The systems would be made up of atoms or their components, and governed by magnetic fields.

The alleged creatures would reproduce magnetically and feed on solar radiation. They would range in size from microscopic patterns of magnetic force created by colliding atomic fragments to electromagnetic fields light-years in diameter.

The physical life-forms on the surface of neutron stars, one scientist suggested recently, would be made up mostly of iron nuclei. Packing about one hundred fifty pounds of weight into an amoeba-sized creature, these beings would have a density seven million times that of water. They would travel along the magnetic fields, glow white because of the heat, and have eyes raised on stalks so as not to blind themselves by the radiance of their bodies. How the eyes would be preserved under such conditions is not explained.

Other scientists have hypothesized about life-forms that live in the interior of stars as patterns of moving charges under the influence of magnetic fields. Others exchange radiation in orderly patterns of laser-like beams.

Some scientists have even suggested that galaxies themselves may be nothing but molecules in huge assemblages of cosmic superorganisms. In this idea, gravitational force would take the place of the electromagnetism that holds our kind of life together. On the neutron stars, on the other hand, the strong force would be the life-force —it holds the atomic nucleus together.

Proponents of gravitational life concede, however, that the universe has not existed long enough for such life to evolve—even if it is possible.

On neutron stars, on the other hand, life-forms would last a mere billionth of a second. Obviously, there wouldn't be any way for us to communicate with such life, or, for that matter, even to detect it, since it might be invisible.

Such ideas obviously cross over into science fiction. When presented as such, and skillfully elaborated, they can make entertaining reading. The noted British astrophysicist Fred Hoyle, for instance, wrote a skillful work of fiction, *The Black Cloud*, in which a giant interstellar cloud, functioning as a gigantic brain, travels through space and informs man of the marvels of the universe when it passes near Earth.

Much like certain marine colonial organisms on Earth, Hoyle's black cloud consists of individual units that specialize in various functions. Electromagnetic currents supply the thinking power to the cloud.

But when such ideas are presented as science, despite their apparent foundation on physics they fall far short of plausibility, because of the extreme simplicity of the proposed life-forms. They suffer from even more serious drawbacks than do silicon and other suggested substitutes for carbon: the level of complexity available in the suggested physical life-forms would be much less than in the simplest carbon system.

Many knowledgeable scientists question the idea of physical (as opposed to biological) life-forms. "Organized magnetic fields, or atoms or molecules lasering in interstellar clouds are certainly terribly simple, and have no reason to be called living," typically says Dr. Jesse L. Greenstein, the noted astrophysicist who recently retired from the California Institute of Technology. "It is certain that many things we call live react to their environment because of simple physical and chemical laws. You could say that a sodium atom eats a chlorine atom to make salt. But we wouldn't believe that a test tube with the reaction making salt going on in it was 'alive.'" The physical life-forms suggested, adds Dr. Greenstein, don't even approach a simple virus in their complexity.

The exception to the second law of thermodynamics—suspension of a general decrease of disorder in nature thanks to the complexity of a system—is highly unlikely to occur in the natural environments described by the proponents of physical life-forms.

"One magnetic field doesn't eat another," explains Dr. Greenstein. "In fact, the organized energy of the electric currents that make magnetic fields is dissipated by atomic collisions and turned into heat eventually."

So, despite occasional disclaimers that what they are suggesting is not science fiction, the authors of such ideas do not contribute much to our understanding or to our quest into the nature of life elsewhere. There is one contribution they do make: to make it apparent that the existence of such life-forms is highly improbable.

Another idea based on physics that always pops up in science fiction and is also expounded by some scientists is a sedentary colonial superbrain. The advent of the electronic computer has revived the popularity of this idea, since suggestions on what is possible in space often reflect advances in our own technology.

In some versions, this superbrain is really a supercomputer that stores all the wisdom and creative achievements of the human race and improves on them. The superbrain doesn't communicate with anyone, only with itself. In other variants, the superbrain becomes an intelligence encompassing the whole surface of a planet, sometimes becoming a part of a larger, interplanetary network.

We have examples of colonial organisms on Earth, some highly intriguing ones. Take the colonies of corals, for instance. Coral reefs are made of limestone secreted by innumerable coral polyps. After the current crop dies, new polyps grow on top of them.

Charles Darwin was the first to suggest that coral polyps develop around the top of a mountain that is subsiding. They build on for millions of years, forming a reef around the top of the mountain, with the living corals situated in the top 150-foot layer of their structure.

The corals propagate by dividing, and while individual polyps are tiny, because they constantly divide, the coral structures grow rapidly.

Adjoining polyps have connecting body walls and gut cavities. To feed, polyps stretch their bodies and tentacles but rapidly withdraw into their limestone houses when danger threatens.

Corals, of course, display no signs of higher intelligence, only reflex reactions.

There are also mobile colonial organisms. These are exemplified by the hydrozoans of the order Siphonophora which faintly resemble jellyfish, some of which are also hydrozoans; with their stinging tentacles, the hydrozoans capture small fish and other creatures. About three hundred species of these peculiar organisms are known.

The Portuguese man-of-war is probably the most familiar of the Siphonophora. On the surface, the man-of-war and the hydrozoans appear to be discrete organisms, but in reality they consist of many specialized member units, or zooids.

Some zooids provide floatation, others supply propulsive power. Still others engage in digestive and reproductive functions. New zooids bud from special growth zones.

In reality, siphonophores are organisms that originated as colonies. Some more primitive types of these creatures have their individual components go through visible changes such as loss of reproductive capacity while the units designated to become reproductive organs lose their ability to feed themselves.

Interestingly, in the Siphonophora *Muggiaea*, the tail-like appendage consists of individual units that can break loose and live independently, later to rejoin the colony. These individuals appear to represent a stage intermediate between a true colony and a fully integrated organism, faintly approaching colonies of social insects, in which individuals, of course, are physically separate and capable of independent motion.

Probably the most interesting point about these strange creatures, however, is the different way in which they created organs out of individual organisms. Other higher animals didn't follow the colonial route; they fashioned their organs out of their own tissue as original equipment.

This is still another example of convergent evolution, showing a different avenue to the same end that may be followed by beings on other worlds.

There are many variations of colonial forms on Earth, suggesting further how colonial organisms may evolve elsewhere. Some parasitic fungi and molds, for instance, form a mass after penetrating the wall of a green plant. Sponges, another colonial organism, consist of semi-independent individuals that constitute the whole. Rotifers attach themselves to the tubes of older individuals to compose a radiant colony that floats through open water as a unit. Sea squirts and free-swimming plankton-feeders known as thaliaceans consist basically of feeding units that open into a common "stomach."

Biologists have also been fascinated with the colonial organization of certain slime molds. They start out as individual amoeba-like creatures that frequently divide. As long as food is plentiful, the single-celled amoebas lead separate lives. But as soon as food becomes scarce, they congregate into colonies that become transformed into fruiting bodies resembling tiny trees. Spores grown and elevated on stalks comprising individual cells can then be dispersed to begin a new search for food.

In no instance, however, have sedentary colonies evolved any intelligence. The idea runs counter to the evolutionary experience. Mobile colonies of insects have done much better in that respect, underscoring the fact that an active, exploratory way of life, not a sedentary one, is what leads to the evolution of a large brain. So the scientists who visualize a computer-like, sedentary superbrain have nothing to lean on in nature.

Another problem with most scientists speculating about possibilities of life outside and beyond Darwinian evolution is that scientists generally are late entrants into the game. Good science fiction writers explored the same territory long ago and usually in much more imaginative and entertaining ways.

A more practical difficulty with unknown life-forms is that even if they exist, it may be difficult, even impossible, to detect and recognize them.

As the continuing controversy over the biological findings of the Mars Viking spacecraft shows, it has been difficult enough to determine whether life as we know it exists on Mars.

The definition of life familiar to us is a fairly arbitrary one. In the main, there is agreement among the scientists that life could be defined by its properties or attributes. The one definition that appears most reasonable is that life is an organization that is made up of macromolecular substances that function together to reproduce their species. That would describe a living substance.

By this definition, would a virus be considered living? Probably not, because while it can provide a blueprint, it cannot function on its own. A complete organism is one that can live independently and produce its own kind. A parasite may live inside a host and the host provide the proper environment for it, but the parasite is still an entire organism, while a virus is merely an activator for the host. Outside the host, a virus is a collection of inert chemicals. The host does all the work. The virus, having invaded the host, simply says, "Get started," and the host makes more of the virus particles. Essentially, the virus is a nonentity as far as the host is concerned. It springs to life only when it finds the proper environment, namely that of the host.

What about crystals in this context? A crystal shows one attribute that is a measure of biological activity: when crystals grow, they form other crystals, thus reproducing themselves. But a crystal doesn't have any of the other properties of life. It doesn't metabolize or create matter out of other matter. It is very much like a virus in that sense. The environment causes the crystal to be there; a crystal can form only in a saturated solution of its kind. Life doesn't occur that way.

NASA had to keep those basic facts in mind when its scientists were designing life-detecting equipment for the Mars landers. The scientists had to remember that there isn't one single property in a biological system that can be used by itself to ascertain the presence of life, that it is important to be able to measure as many of the attributes as possible within a given sample.

But what signs would we look for in order to detect life of an unknown type?

Would we be misled into believing, for instance, that silicon life was possible if we found fossils of the tiny unicellular organisms called Radiolaria, which compose part of the floating plankton of warm seas? The Radiolaria extract highly diluted silica from the seawater, from which they fashion strikingly beautiful protective and supporting structures.

These tiny siliceous skeletons resemble glass, and some of them look somewhat like snowflakes while others resemble miniature helmets, latticed balls, and even tiny sunflowers.

If Radiolaria did not exist on Earth and if we found siliceous skeletons of similar creatures on Mars without knowing anything about the creatures that had lived inside and around these structures, would this be hailed as the discovery of silicon-based life? Chances are good that it would.

But Radiolaria's use of silica, of course, is strictly structural. No living thing on Earth uses silicon, or any of its variations, as a functional molecule.

Are the possibilities, then, confined largely to variations on the terrestrial theme of life?

All signs point to that conclusion, with a single exception to which we'll come in a moment.

Life based on carbon is favored as the dominant kind of life in the universe because, as we have seen, the abundances of various elements appear to be approximately the same everywhere in the universe as measured from Earth. In fact, carbon, oxygen, hydrogen, and nitrogen, the four principal elements from which our kind of life is constructed, constitute 95 percent of the matter in the universe. Carbon-based life can thus be expected to be the rule, because it mirrors the nature of the universe.

Therefore, we are not being too narrow-minded when we refuse to seriously consider the existence of physical or other life-forms that don't make biological or common sense. When best materials are on hand to construct life, why expect nature to choose second-, or even third-grade candidates, or go to fire and brimstone, instead of placing life in more suitable environments?

Does this make the whole subject of extraterrestrial life less exciting?

On the contrary, it makes it more so. For the conclusion that life as we know it is the dominant kind greatly increases the possibility that if and when we contact or encounter creatures from other planets, they will be much like us in terms of makeup even if not our duplicates in exact shape. They are not going to be some sulfur-breathing monsters with ammonia flowing in their veins or some strange filaments or whirlpools of energy.

At least, the creatures elsewhere will be much like our own kind on planets that approximate the age of the Earth. Age is the key word, because perhaps the most difficult concept for us to grasp and to try to visualize is how the passage of untold millions and even billions of years will remold living beings.

We have been around for such a short time, yet we have changed so much. Will animals on planets billions of years older than the Earth differ as much from our kind as the beautifully caparisoned butterflies differ from the dumpy caterpillars from which they come?

Our next chapter will try to answer that question, at least as it applies to the future of man.

CHAPTER EIGHT

Future Man

The big imponderable in evolution is how the passage of untold millions and billions of years will remold creatures on Earth and elsewhere—or, for that matter, whether it will leave them largely unchanged. Some animals, apparently fitting their niches to near perfection, have remained unaltered for hundreds of millions of years: frogs and turtles are two examples. Man, obviously, is much younger than that and probably is still changing. It's doubtful, in any case, that he has reached a state of perfection, at least not in his behavior, which, like all his other activities, as well as his bodily structures, is governed by his biological past. But what lies in man's future?

It's a question that can be approached from at least two directions: by considering what has happened to the evolution of intelligence on Earth, and what could have happened on planets of other stars.

Where our cosmic counterparts are concerned, it is fashionable for the physical scientists to speculate that other intelligent beings in space have gone unimaginably farther than we have in development. In support of that notion, they cite the fact that most stars in the universe are much older than the sun. So that if life-bearing planets circle those stars, according to this argument, life there would be billions of years older than on Earth and could have evolved into forms that would be difficult for us to even visualize.

This idea assumes a straight-line, uninterrupted advance in the development of intelligent creatures. It's a somewhat naïve idea, because the evolutionary record on Earth does not support it. Our fossil record is obviously incomplete, but the geological record even today does not yield what David M. Raup, dean of science at the Field Museum of Natural History, in Chicago, calls "a finely graduated chain of slow and progressive evolution." What is visible in the fossil record, instead, is a lot of unevenness, a rather sudden appearance of a new species and its equally sudden disappearance, to be followed by species that are not necessarily a biological improvement. This does not imply, of course, a creationist method of evolution but, rather, shows the jerky, fit-and-starts probing of the branches of the evolutionary tree.

So it appears with the evolution of intelligence. There is a common belief that biological systems, once initiated, tend to evolve toward higher levels of complexity and that such complexity naturally leads to intelligent life. But intelligence is not a product of a general trend in evolution, any more than is any other attribute of life. Evolutionary progress is not determined by some ultimate goal, and that goes for the evolution of intelligence, too. Instead, as Simpson has noted in *Life* (Harcourt, Brace & World, 1965), only one universal trend characterizes evolution: "a tendency of life to expand, to fill in all the available spaces in the livable environments, including those created by the process of that expansion itself."

Intelligence, therefore, appears to evolve not because of some mysterious trend toward greater complexity but because animals take advantage of environmental opportunities. And at that, the development of intelligence comes about strictly by accident.

Thus, attainment of the upright stance by man's predecessors freed their forelimbs for many new tasks: carrying objects, toolmaking, and other complex movements. The freeing of the forelimbs stimulated the development of the brain. Our big brains, as well as our ability to speak, thus are a result of specialization in the search for food and in its consumption. The new diet modified our ancestors' jaws and oral cavities, enabling them to speak. Man emerged not only as an intelligent being but also as a cognitive one.

The distinction between intelligence and cognition is all-important to our understanding whether or not creatures similar to man in brainpower exist elsewhere. We owe the delineation of that distinction to C. Owen Lovejoy, of Kent State University. If intelligence is defined as a capacity to comprehend relationships, all animals with the ability to store and use information that has a bearing on relationships can then be called intelligent. Cognition, on the other hand, is a higher form of intelligence, an ability to use symbols (made possible by speech), to formulate abstractions, and to recognize the self.

By this definition, man is not only an intelligent but also a cognitive animal—probably the only such being on Earth, unless dolphins are cognitive too. He may also be one of the few cognitive beings in the universe, because cognition is not the result of an evolutionary trend but the end product of many unrelated evolutionary adjustments that, luckily for us, all added up in a cognitive being called man. Most of the population of the Darwinian universe, on the other hand, would be placed in the intelligent—but not cognitive—category. Cognitive beings may be exceedingly rare in the universe, while intelligent ones may be extremely common.

The accidental nature of the evolution of man's brain, furthermore, might have been influenced by geological upheavals that set the particular rate of evolution of intelligence on Earth that led to man.

This new idea was proposed recently by Dale A. Russell, who heads the paleobiology division of the National Museum of Natural Sciences, in Ottawa, Ontario. It's an idea that could radically change man's status along the scale of intelligent life in the universe.

Dr. Russell figures that an understanding of the factors that can affect the evolution of intelligence on Earth is of great importance in evaluating the cosmos as a home for intelligent organisms. The creatures now living on Earth, as well as the fossil record, must reflect the effects of these factors, Dr. Russell feels. He knows, of course, that the existing body of relevant information is both imprecise and incomplete. Accordingly, Russell concedes that his suggestions contain a large speculative element, but he also feels that this whole new arena of evolution of intelligence is ripe for exploitation by scientists of differing disciplines.

What he has found is support from geology for the view that evolution of intelligence on Earth has not proceeded in a straight line. The highest levels of intelligence attained by organisms through geologic time, Russell has concluded, follow a pattern that suggests that environmental conditions affect the rate of increase in intelligence. In light of the all-pervading influence of the environment on the development of life, this finding is not surprising. (See Chapter Five, "Dictates of the Environment: The Look of Life.")

What is surprising, though, is Dale Russell's conclusion that evolution of intelligence on Earth occurred at two strikingly different rates. First, evolution of intelligence proceeded fairly rapidly during the time 700 million to 200 million years ago. During the early part of that period, life was confined to the oceans. If this older, faster rate had continued, man would have attained his present brain size 60 million years ago. Or, at least, *some* kind of creature would have developed a brain that size, since mammals were just emerging then.

Something happened, however, about 200 million years ago to drastically slow down the overall rate of brain development. Russell speculates that the difference in the brain growth rates during the two geological periods may be accounted for by the cutoff of the flow of terrestrial nutrients into the oceans as life established itself on land. Conditions for the growth of brains of land animals improved, but the consequent malnourishment of ocean-dwelling organisms resulted in a net lowering of the rates of maximum increase in brain size of creatures populating the planet as a whole.

A much slower rate of brain development set in—a rate at which human intelligence could not have evolved for an astounding 20 billion years if the more recent rate had prevailed since the beginning of life on Earth. If that is the "normal" rate for evolution of high intelligence everywhere, a striking conclusion of Russell's research is that man is the most intelligent being in the universe right now—a sobering thought. Another consequence of this line of reasoning would be a conclusion that the origin of life is intrinsically a much more probable event than the origin of higher intelligence. What we don't know, of course, is which rate of brain evolution prevails elsewhere. But we can guess that differing rates are at work on various planets, so that man is likely to have both younger and older intelligent cousins.

Where man is concerned, the average rate of his brain development has doubled during the past two million years, and the trend toward improvement of brain function should continue, Russell suggests. In fact, he projects that if the trends established in the past go on, nine hundred thousand years from now the human brain will have expanded to three times its present size, and the human life-span will have doubled. This new man would have a much thicker neck to support the additional weight of the head, without increasing total body weight. Babies would be born after a shorter term and in a more immature state neurologically so the fetus's head could still pass through the birth canal.

"When human brain size doubled during the last two million years," says Russell, "man acquired language, libraries and a technology which has enabled him to explore the planets. The projected increase in brain size could not unreasonably be postulated to produce an analogous expansion in human mental abilities." It would place man of the future as far ahead of us as we are of the apes.

Such a leap in mental powers, Russell thinks, would enable man to sever the link that ties his fate to that of the sun; the future humans could escape the sun's expanding, red-giant phase by building self-sustaining space habitats farther out in the solar system.

Not all scientists agree, to be sure, that human intelligence is improving or that evolution of intelligence will continue.

Extrapolating evolutionary trends into the future, as George Gaylord Simpson has noted, involves the realization that evolutionary trends are governed by natural selection. As long as they last—and they don't continue indefinitely—evolutionary trends have adaptation as their basis. But the trends can stop, go off in another direction, or even reverse themselves.

Simpson, for one, has questioned the popular assumption that because man's brain has greatly increased in size it will continue to do so. He has argued that it is unlikely that such a trend now exists or that it can be shown to have existed in man's recent history. Man's direct forerunner, Cro-Magnon man, for instance, had a brain similar in size to ours. The still older Neanderthals, strangely enough, apparently had brains slightly larger than ours, a fact that puzzled Darwin. Neanderthals are held by most scientists to have been a side branch on the evolutionary tree, a branch that died out. Although not everyone agrees with that interpretation, the idea that increased brain size is not necessarily being favored today by natural selection is held by a number of scientists.

There is little doubt, however, that the changes that confront man today are cultural, affecting what he does, rather than what he is. For biological evolution, man has successfully substituted the cultural and technological ones. He has expanded his abilities from running and climbing to flying and even journeying through nearby space. He has built clever instruments to probe the depths of the atom and of the cosmos. He has extended his communication capabilities to span continents and oceans and even the gulf of space. Through cultural accessories to his body and brain, man has become the most adaptable of all species—and ironically, the most dangerous to himself. Even more ironically, by those accomplishments of his brain, man may have largely eliminated the pressures that otherwise would have served to foster further biological evolution. At least, some scientists think so. Most biologists hold, however, that man's evolution is continuing and will continue, even though it may have slowed down. Furthermore, biology and culture are bound to interact, with cultural factors possibly influencing the nature and the rate of biological evolution.

Thus the potential for natural selection most likely continues to exist even in advanced technological societies. But scientists find it hard to determine what is actually happening. Where natural selection is concerned, they know that such factors as prenatal and postnatal death, as well as fecundity, all contribute to the evolutionary potential. They also know that, in general, natural selection favors physical fitness, although man's increasing ability to allow people with "defective" genes, as expressed in various genetic diseases, to live out all or most of their normal life-spans and to reproduce, may eventually reduce the total fitness of the race.

There are ways to prevent possible genetic degeneration of the human species, to be sure. Genetic counseling, in which people with defective genes are advised about the chances of the defect coming to the fore in their children, is an example of such prevention already at work. More advanced techniques are now on the horizon, such as intervention into genetic defects with the emerging tools of genetic engineering.

The advent of molecular biology, in fact, may fuel the decisive change in man's biological future. Even now, we already understand the basics of cell organization and reproduction. Gradually, the full details of the chemical basis of life are being unraveled. The key processes of cell development and differentiation are likely to be understood in the next fifty to one hundred years. There is no question that this knowledge will be practically applied; we already have a nascent genetic engineering industry that can produce biological substances in short supply. From there, it isn't such a big step to growing spare genes and whole organs to replace failing ones in order to correct defects in man's physiology that leave him open to disease and the deterioration of old age.

A number of avenues into which the new biotechnology can be channeled are already visible. We are approaching the time, for instance, when it will be possible to read the sequence of DNA bases in the human sperm or egg cell, analyze the genes with a computer to detect deleterious ones, and then replace them with good genes via microsurgery. New genes to impart desired characteristics to an individual might be added.

Technical feasibility aside, such tinkering with human genetic material obviously would raise ethical problems of immense magnitude and severity. It brings to mind the worst nightmares of dictatorial regimes breeding Nazi-like "supermen," and the racism of the eugenics movement.

But it is remarkable how rapidly at least some aspects of human intervention into the planning of progeny can become accepted. As recently as thirty years ago, for instance, test-tube babies would have been confined to the pages of science fiction, yet today they are a reality. The possibility of cloning people, or producing exact copies of human beings, also looms in the near future. Already, not only frogs, which, of course, are amphibians and whose large eggs are fairly easy to work with, but also mice, which are mammals and have eggs as small as those of humans, have been cloned. Cloning won't necessarily lead to any improvement in the human race, to be sure, because it reduces genetic variability instead of increasing it. Whether cloning exceptionally bright people will transfer their brilliance to the clones is also unknown. No one can say whether a clone of Einstein would have been another genius or merely a shell of the original.

Test-tube babies, on the face of it, offer a greater opportunity for development of people with larger brains, since they will have circumvented the need to have heads small enough to pass through the birth canal.

Natural selection may need such nudging from man, because, contrary to what most people believe, natural selection in reality does not hasten evolutionary change but slows it down. Natural selection is interested in making an organism—and a species—fit its existing environment (not some unknown future setting) in the best possible manner.

When changes are slow in the natural environment, evolution keeps up with them, skillfully adapting a species to the changes. But when such changes come too fast, as apparently happened when dinosaurs ruled the Earth, evolution can't react fast enough, and a species, or a whole number of them, perish. Evolution, in other words, is not a strategic planner with a clear vision of the future, but a rather blind tactical "claims adjuster," a single-minded modifier concerned with the immediate present. That's why Harvard zoologist Ernst Mayr has noted that "the theory of natural selection can describe and explain phenomena with considerable precision but cannot make reliable predictions."

One way the problem of how to evolve a large brain might have been solved naturally would have been for a man-like creature to evolve from a marsupial ancestor. Marsupials and placentals are believed to have originated at about the same time, about 100 million years ago. Both stemmed, it appears, from the pantotheres, primitive reptilian predecessors of mammals. In the beginning, marsupials outnumbered the placentals, and although placental reproduction is believed to be an advance over the marsupial method, marsupials may well have prevailed, as they did in Australia. As we have seen in Chapter Four, "The Turns and Twists of Evolution on Earth," a menagerie of marsupials—no fewer than 170 different animals—ranging from a marsupial "wolf" to the koala bear evolved in Australia from a single marsupial ancestor, a small, shrew-like creature. Such a turn of events on other continents could have led to marsupial man on Earth.

Given more time, a koala conceivably could have gone on to develop into a somewhat man-like creature. Its females would have given the koala an immense advantage over man. Female marsupials do not possess a well-developed womb and placenta, that lifeline of the fetus. Accordingly, fetuses of marsupials spend only a short time in internal growth. Their mothers give what for placental animals would be fatally premature births. The underdeveloped young then crawl into the mother's pouch, where they attach themselves to the nipples and remain there for months.

While this may look like a primitive means of reproduction com-

pared with placental animals, for a highly intelligent creature the
marsupial way of birth would be a fantastic gift. The fairly narrow
birth canal of human females may block the further evolution of
man's brain. At birth, our heads are already so large that they barely
fit through the pelvic opening. If the brain is to grow further, future
man must either enlarge the birth canal or manipulate the developmental processes so that the brain can continue to grow long after
birth. Our marsupial cousins, though, could grow much larger heads,
with brains like those Dale Russell projects for man nine hundred
thousand years from now. The marsupial female facing this page
may strike us at first as very odd, but her children may be unimaginably ahead of us in intelligence and technology. (See color insert.)

The remarkable phenomenon of evolutionary convergence, which
crosses the barriers of space and time, would allow the emergence of
marsupial man elsewhere, assuming there was a predecessor sequence
of fish–amphibian–reptile–mammal. Man's evolution, as we have seen,
is built on such a collection of improbable events that to meet an
exact duplicate of man in the cosmos would be as unlikely as to encounter a double of a kangaroo that had evolved in Alaska.

Yet Alaska has the caribou—its counterpart of the kangaroo.

So the possibility of our having marsupial cousins in the cosmos
cannot be completely excluded. They would look more like koala
bears than men, and they may, or may not, build technological civilizations. Man's speculation about other civilizations in space is based
on the assumption that intelligent beings in the cosmos will go from
one invention to another, just as we did—from the wheel of the horse
carriage to the giant radio telescope dishes that now wheel across the
skies. The assumption may be faulty, of course, because nontechnological societies are possible too.

As for man's future evolution, at least the natural part of it is more
likely than not to continue along the lines laid down in the past. And
what a checkered past it has been! In our blood we carry the salt of
primordial oceans. Our arms and legs were once fish flippers. Our
women's amniotic sacs link them directly to the reptiles' amniotic
eggs—and for both men and women such links to reptiles extend
through the reptilian portion of our brain, that knob-like structure
that constitutes the brains of crocodiles, frogs, and lizards today. To
be sure, the reptilian brain has been wrapped by the later, mammalian brain and offset by the neocortex, which makes us human, or al-

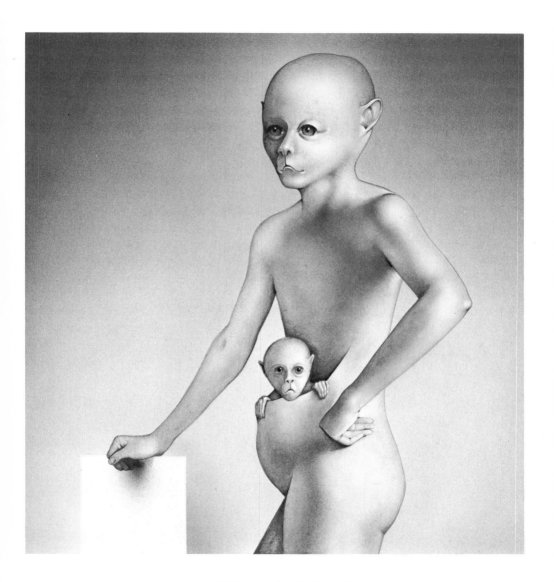

Given the many possibilities of evolution, it's not inconceivable that under certain circumstances, a marsupial could evolve into a highly intelligent biped.

most. The ancient reptile brain still is very influential; it mediates our emotional behavior and has poor communication with the higher, newer brain of reason. As Arthur Koestler once poetically put it: "There seems to be an icebox built somewhere into our skulls which is the seat of cold logic and rational thought, and next to it there is a hothouse of irrational, mood-based beliefs." Scientists have traced all sorts of problems to that lowly reptilian brain from whose hold man has been unable to escape. How else do we explain cruelties both ancient and modern that man has perpetrated against his fellow man?

Elsewhere, intelligent beings may have escaped this cruel legacy by not having had reptiles as ancestors; the octopus, for instance, is one such creature. It does not display the cruelty associated with reptiles. For man, though, the outcome of the continuing struggle between the reptile and the human in his head could prove decisive to his future. We may be smugly pleased with man as a supposed end product of evolution, but we shouldn't forget that most of nature's evolutionary experiments wind up in a dead end. It was believed at one time that dinosaurs' puny brains caused their downfall. Now man's large brain threatens his extinction. Does intelligence, then, offer an advantage for survival? It appears that the answer depends on the nature of the intelligence. And it may well be that unless man somehow brings the reptile in him under control, enlarging man's brain would only make him a more skilled warrior and schemer against fellow men. We should remember that the average duration of a species on the Earth is less than 10 million years and that within the 600-million-year record of really abundant life, a biological turnover has changed the cast of animal actors many times. So the hope of man's survival lies in his brain too.

As for man's body, considering its history the body is as good as it can be. It changed from a fish into a four-legged terrestrial walker, a four-legged tree swinger, and then into a two-legged runner, erect and with a torso balancing on feet and hips, and the head on the base of the neck. Alongside its advantages, the upright stance created some problems, such as narrowing of the pelvic opening in women, thus complicating the birth process because of the human fetus's large head. The elevation of the body, furthermore, has shifted its support from four corners to two, putting a lot of weight on the pelvis, causing slipped discs and similar problems. And the human foot, consisting of a collection of about forty bones and remade three times, from a flipper into a climbing foot and then a walking foot, is not really designed for walking flat-footed on cement and asphalt, in a confining and often ill-fitting shoe. It is designed, rather, for walking on grass, soft soil, and sandy beaches. So there is room for improvement where the human body is concerned.

There are scientists who argue, to be sure, that the material state of man is only temporary. "A billion years ago, man was a worm in his ancestry," says Robert Jastrow. "A billion years from now we will have evolved as far from our present form as we have from a worm."

"I don't think that life such as ours with a lot of water and the carbon chain is more than an ephemeral stage. Knowing the length of the universe's existence and the short time the Earth has existed, I don't think that life in the cosmos, more than a minute fraction of it, is some distorted replica of our chemistry. I think it's either disembodied life of mind entirely, or else in the silicon form—not the sand-eating monsters—but what we call the computer. I don't think there will be a carbon-based bag of blood-and-guts with our shell of bones housing intelligence—they just won't have any biological containers whatsoever. These mortal frames that last a hundred years must be a very primitive model in the cosmos. I would put my money on the silicon memory bank as an immortal form of life and on the disembodied form as the ultimate residence of mind.

"My belief is that every planet which has suitable water and temperature goes through four or five billion years of carbon evolution, because as Ponnamperuma says the ingredients are abundant and the reactions go well and the Urey-Miller experiment showed that everything starts off nicely.

"And then I think after four or five billion years life passes on to a less vulnerable and more expandable framework for housing its intelligence. And it leaves the carbon-based life behind it. So there are creatures like us around but they must be very thinly scattered and not the majority because we intelligent humans are a very thin slice of time."

The idea that man somehow will evolve into a disembodied intelligence, a kind of God if you will, mirrors the mystical views of such religious thinkers as Father Teilhard de Chardin, who foresaw a highly directed course of evolution. Man, in that view, was only evolution's intermediate goal, to be followed by a change into a God whom Teilhard called Omega. The course of evolution on Earth offers no support for these ideas, however.

Man's biological future, instead, is imbedded in his past. To be

sure, the new element that may change all this is man's emerging ability to direct his biological destiny. Deliberate restructuring of his brain and body may allow man to ascend into a luminous future up the DNA ladder, as shown in the illustration facing this page.

Have intelligent beings elsewhere already resorted to such genetic improvement? We may soon find out. Scientists are hopeful that evolutionary convergence works in brain structure and ways of thinking just as it works in other organs, such as the eye and the hand, so that we could communicate with such alien intelligences. All brains presumably have to deal with concepts—arbitrary sets of symbols—and to form certain kinds of associations. But first we have to locate other life-bearing planets and their inhabitants. The next chapter reveals how surprisingly close we may be to that goal.

Part III

THE SEARCH NARROWS

CHAPTER NINE

Planet Search

Lost in the glare of distant suns like specks of dust in faraway whirlwinds, billions of planets bearing a fantastic diversity of life circle other stars, unseen by our instruments.

Unseen, but not for long. At long last, we now stand on the threshold of one of the most exciting events in the history of science, for within this decade those mystery worlds will be a mystery no more. Highly advanced new detection techniques, now under rapid development, will find and probe those planets in a variety of ways. They will photograph them and take their temperatures. They will tell us by the color of the planets not only whether they have life-generating oceans and green plants but even disclose what their atmospheres are made of. They will measure the diameters of the planets and their orbits—all thanks to the explosive capabilities of modern electronics. Little wonder that David C. Black, an astronomer at NASA's Ames Research Center, at Mountain View, California, and one of the leaders of the planet search, feels that "this is an exciting time to be in astronomy."

The existence of planets outside our solar system has been suspected for some time now, of course. Astronomical evidence, albeit
indirect, and theory, as well as plain logic, all point in that direction.
Where logic is concerned, the Chinese philosopher Teng Mu expressed the idea poetically in A.D. 1200 by saying: "How unreasonable it would be to suppose that besides the Earth and the sky
which we can see, there are no other skies and no other Earths."
Where astronomical evidence is concerned, some astronomers have
been in fact convinced for decades that they had already detected
Jupiter-sized planets tugging at at least one parent star.

The most widely publicized case involved the supposed discovery
of Jupiter- and Saturn-sized planets orbiting Barnard's star, a faint
red dwarf that is among the stars nearest to us. This discovery has
now been seriously questioned. It now appears that Barnard's star
never had a planetary system, at least not of the type reported.

That loss will soon be remedied, however, by what's coming; the
new instruments offer unprecedented accuracy, which will settle the
question of planets elsewhere once and for all.

Unequivocal confirmation of the existence of other planetary systems, along with a discovery of signs of life on those planets, would
do more than excite public imagination. It would finally put the
whole science of exobiology, or as the Russians call it, more fittingly,
astrobiology, on a much more solid footing. No longer could clever
critics remark flippantly, as one did, that exobiology "is a science
whose subject matter doesn't exist."

Furthermore, confirmation that life-bearing planets are common
would fire the enthusiasm of scientists and the public to learn more
about the distant worlds. Such a discovery should induce Congress to
appropriate funds to mount a substantial effort in the search for extraterrestrial life—an effort that is still on a shoestring budget today
despite its universal appeal and great importance.

There is little doubt among the scientists involved in the search
that those critics and the rest of us don't have long to wait for the
big payoff.

But to put the imminent leap in planetary detection in a historical
context, we must glance briefly at the scientific underpinnings of the
search.

Ideas on whether planetary systems such as ours are common or
not have swung like a pendulum in the past two centuries. As a result

of the work of Laplace and Kant, at the beginning of the nineteenth century, many scientists believed that planets formed as a result of a solar nebula breaking into rings. The idea that there must be planetary systems elsewhere was widely held. By 1900, however, unable to explain why the sun has such a low spin rate, or angular momentum, compared to the planets, many scientists began to embrace catastrophe theories, which would make planetary systems uncommon.

In the 1950s, though, the pendulum swung again. Now scientists feel that they can account for the loss of angular momentum by the sun through magnetic fields and solar winds. This brought back the evolutionary ideas prevalent today: the idea that formation of planetary systems is a natural if not inevitable by-product of star formation, a process "as normal as childbirth," as one American astronomer has called it.

What happens, according to this hypothesis, is that when the gas clouds that form stars condense, they apparently break up into a number of pieces, forming systems that consist of two, three, four, and even more stars orbiting a common center of gravity. Observational evidence, in fact, confirms this idea: most stars are part of such binary, or double, or multiple systems.

The distinction between multiple star systems and our planetary system, in the main, appears to be one of size. At that, it isn't such a large difference. Stars in an average binary system are situated closer together than are our sun and our outermost planet, Pluto. The distance between average binary stars is two billion miles, while Pluto is twice that far from the sun. This has led many scientists to conclude that formation of companion stars and planets represents a continuum, a process that gives birth sometimes to stars and sometimes to planets.

In many cases, the separate components of binary and multiple stars are readily visible, sometimes even with the naked eye, but more often through a telescope. Sirius, the Dog Star, for instance, is accompanied by its "pup," a faint white dwarf. The stellar family known as Castor, in the constellation of Gemini, on the other hand, consists of six stars, four bright and two faint. And the three stars closest to the Earth, Alpha Centauri, with its components, A and B, and their more distant companion, Proxima Centauri, constitute a three-member family, all rotating around a common center of gravity.

Because some of these companion stars are situated no farther from their primary stars than Neptune is from the sun, it's logical to ask whether planetary systems like ours wouldn't form in some cases instead of stars.

Indirect evidence, indeed, points to the widespread existence of planetary systems. Astronomical observations, for instance, show the presence of silicate and probably icy raw materials as dust around newly forming stars. Theoretical models, furthermore, suggest that this material could aggregate efficiently.

What's more, astronomers have determined that there is enough silicate in every star in the galaxy to produce not one but several Earth-like planets near each star.

Most of these planets, furthermore, would be much older than ours. Probably 90 percent of all the stars in the universe are reddish dwarfs with surface temperatures as low as 4,500 degrees Fahrenheit (2,500 degrees C), compared with the sun's 10,000 degrees F (5,500 degrees C). The red dwarfs emit only about one hundredth the sun's radiation, and that mostly in the infrared. Because they use up their nuclear fuel so slowly, these stars are incredibly long-lived.

Many of them may have already existed when our sun, a relative newcomer among stars, lit up, five billion years ago. That would bring some of the red dwarfs close to the age of the universe, about 15 to 20 billion years. And they will continue to shine for tens of billions of years to come, long after the sun has exhausted its fuel in a fiery fling that will make it bloat into a red giant that will burn up the Earth and other planets near this nuclear furnace in the sky run amok.

In terms of planets and longevity of life on them, the red dwarfs obviously offer a more permanent abode. And intelligent life on such planets will, of course, have moved unimaginably beyond ours.

It was at one of these red dwarfs, Barnard's star, that on a fateful day in 1956 the tall and courtly Dutch-born American astronomer Peter van de Kamp aimed the long tube of the Sproul Observatory's 61-inch refracting telescope.

Van de Kamp's interest was fueled by an interesting discovery by Otto Struve, the Russian-American astronomer who found that sun-like stars rotate very slowly. To Struve this suggested the intriguing idea that the angular momentum of those stars had been transferred to planetary systems, just as the angular momentum has been removed from the sun by its planets. If true, this discovery would make planetary systems extremely common.

No matter how intriguing the indirect evidence for the existence of other planetary systems, obviously it would be much more satisfying to observe such planets more directly. Such evidence could not have been obtained by direct photography, since planets elsewhere would be hidden in the glare of their suns. In visible light, for example, our sun outshines Jupiter by a blinding two and a half *billion* times. There was thus no way then to try to photograph a planet directly.

Van de Kamp knew this, of course, and decided that astrometry might do the trick. Astrometry is the time-honored technique that allows the observation of the relative motions of stars and planets. In fact, it was astrometry that originally gave birth to astronomy. In ancient times, astronomy consisted basically of astrometry, which was employed to follow the motions of stars and planets to predict the best time for planting crops and other practical activities. In more modern times, as technology expanded astronomy's capabilities, as astronomers concentrated on more exotic objects such as quasars and pulsars and began to explore new wavelength regions such as X rays, radio, and ultraviolet, astrometry became something of a stepchild.

The development of astronomical instruments reflected this trend. To see farther into space, to discover exciting new objects, telescopes were designed for maximum light-gathering with big mirror surfaces, rather than for the highly precise astrometrical measurements that could be used to detect extrasolar planets.

Still, by the early decades of this century, astrometric techniques had improved to the point where it became possible to measure the separation between two photographically recorded star images with accuracies of a few thousandths of a millimeter.

Dr. van de Kamp felt, therefore, that he could try to detect planet-sized bodies with the Sproul Observatory's telescope. He and his associates, after all, had successfully pioneered in the detection of unseen companions of binary and multiple stars. This work is described by other astronomers even today as having been superb. It involves some tricky star tracking.

When a star is accompanied by companions, even if they are invisible because they are too faint or completely dead stars (black dwarfs that have exhausted their nuclear fuel), such companions can still be detected, because they introduce a perturbation, or wiggle, in the basic proper motion of the star through space. Without companions, the star's motion would appear as a straight line. But when nearby companions tug at the star, the line of its motion, when viewed from a side, assumes the shape of a rollercoaster. Our moon imparts a similar wiggly motion to the Earth, and our planets to the sun.

Relatively massive stars, often too faint to be seen by all but the most powerful telescopes, in the intermediate range between stars and planets, were already known to circle other stars in close proximity. They had been detected by astrometry. Van de Kamp thought that by pushing the capability of the Sproul telescope to its limit, he could detect planets circling nearby stars. His choice fell on Barnard's star because, at six light-years away, it is the closest star as seen from the northern hemisphere. It is named after its discoverer, the American astronomer Edward E. Barnard.

Actually, Sproul Observatory already had nearly twenty-five hundred photographic plates on hand, taken since 1938, on which Barnard's star appeared. The Sproul astronomers had almost become convinced that Barnard's star had a small companion. But they kept quiet about it.

The Sproul astronomers, in fact, thought that they had detected dark companions around at least five nearby stars. Their suspected planetary systems displayed the following parameters:

NAME OF SYSTEM	DISTANCE FROM EARTH IN LIGHT-YEARS	PLANET MASSES (TIMES JUPITER)	ORBITAL PERIODS IN YEARS
Barnard's Star	5.9	1.1	26
Lalande 21185	8.2	10	8
Epsilon Eridani	10.8	6–50	25
61 Cygni	11	8	4.8
BD+43°4305	16.9	10–30	28.5

Of these, Barnard's star appeared to have the most planet-like companion—the others could have been black dwarfs. Van de Kamp set out to make certain.

Sure enough, the effort seemed to pay off spectacularly when, only seven years later, in 1963, Van de Kamp announced to a startled world that his data indicated the discovery of a planet slightly larger than Jupiter circling Barnard's star. The planet supposedly orbited the star once every twenty-six years, compared with Jupiter's twelve. In 1969, Van de Kamp added to the excitement when he reported that the perturbations in the orbit of Barnard's star also showed the presence of a Saturn-sized planet.

These surprising discoveries were hailed by newspapers everywhere and found a seemingly permanent place in both popular and professional books on astronomy. "Another Solar System Is Found 36 Trillion Miles from the Sun," read the headline in the New York *Times* on April 19, 1963. *The Cambridge Encyclopedia of Astronomy* declared as recently as 1977: "These are indeed planets! . . . One of the exciting aspects of this discovery is that evidence of a planetary system has been found in a very small sample of nearby stars where an effect could be detected." Encouraged by the apparent proof of planets elsewhere, a group of British engineers and space enthusiasts conducted a detailed paper study on what it would take to build a nuclear-powered spaceship to send on a flyby photographic mission to Barnard's star's planetary system. (See Chapter Ten, "Contact!").

And then the heavenly roof fell in. Unfortunately, recent studies with newer telescopes show no perturbations of the type Van de Kamp described. The wiggles apparently originated in the Sproul refractor. Neither have the new studies confirmed the presence of the larger dark bodies around the other stars suspected by Sproul astronomers of having been planets.

Astronomers elsewhere had noted bothersome similarities in the results applying to all those proposed planetary bodies. Not only were the supposed perturbations always at the error edge of the telescope, but the orbital periods of the proposed planets discovered with the Sproul telescope were all very nearly multiples of eight. The 61 Cygni study was an exception, because it was only partially based on data from the Sproul telescope.

Adding to the other astronomers' concerns, the eccentricities of the calculated planetary orbits were all extremely large, a highly unusual phenomenon, since the orbits of most planets of the solar system are exceedingly circular.

These new findings have pretty much eliminated all those supposed planetary possibilities. Instabilities in the Sproul-telescope optics apparently led to the spurious results.

The fall of Barnard's star was the hardest, since its companion appeared most planet-like.

Raymond P. Vito, professor of engineering at Georgia Tech, even commented on these travails with this "Ode to Apodization":

> Twinkle, twinkle, little star
> Thirty parsecs from where we are.
> Does your wobble through the sky
> Mean a planet is nearby?
> Or has the result come into being
> Because of one arcsecond seeing?*

Most astronomers now feel that Barnard's star has no companions of the size suggested by Van de Kamp, but some scientists, particularly those directly involved, are keeping the issue alive. Now retired and living in Swarthmore, Pennsylvania, where Sproul Observatory is situated, on the Swarthmore College campus, Van de Kamp still believes that his measurements were accurate. So do some of his colleagues. One of them followed a speaker who presented data denying Barnard's star planets, at the meeting of the International Astronomical Union in Montreal in 1980. Van de Kamp's friend argued that the planets do indeed exist.

After the second speaker had finished, a baffled Iosif Shklovsky, a noted Soviet astrophysicist, asked Caltech's Jesse Greenstein, who presided over the session: "Professor Greenstein, did I hear one speaker say that Barnard's star has planets and another argue that it doesn't?"

Greenstein jokingly replied that the two clashing stands represented "an internal contradiction in capitalist science that can only be resolved by dialectical materialism." Shklovsky broke up laughing.

Actually, the contradiction will soon be resolved by a whole range of improved and new instruments that are being developed in the United States to open a new chapter in planetary detection. Barnard's star may still have planets, to be sure, possibly smaller than the

---

* Apodization is an optical technique that improves telescopic vision; a parsec is a measure of astronomical distance equal to 3.2 light-years; arcsecond seeing refers to astrometric accuracy, not sufficient in this case to tell if a star has a planet.

ones Van de Kamp thought he had detected. The new instruments will answer that question. Eventually, the sharpest-eyed of the new generation of planet detectors should be able to spot planets the size of the Earth and smaller around many visible stars in our galaxy.

The magnitude of this kind of advance, which will be achieved in this decade, startles even some scientists involved in the work. In its totality, the advent of these new instruments will equal the arrival of the electron microscope, which allowed us for the first time to individually see such previously unseen minute but vitally important objects as viruses, millions of which could fit into the period at the end of this sentence.

So it will be with extrasolar planets. If the scientists are lucky, they may actually be able to photograph such planets as barely visible dots. Direct photography of planets elsewhere will be difficult, and this achievement may not come for a decade, although surprises are possible.

But there are more readily available, although less direct, ways to look for planets outside our solar system. Improved astrometry and related methods will be used to put stars suspected of having planets under the new kind of astronomic scrutiny.

NASA is now accelerating such an extrasolar-planet-detection program. In the past few years, the space agency has periodically assembled some of this country's leading astronomers in what it appropriately calls Project Orion, since *orion* is the Greek word for hunter.

The aim is to design instruments that will permit unambiguous detection of objects with the mass of Jupiter circling nearby stars. Planets could be larger, of course. The upper limit of the mass of a planet is set by the lowest mass consistent with the definition of a star that shines by nuclear fusion: about six hundredths the mass of the sun.

The planet searchers want to take advantage of the significant advances in electronics that have taken place in recent years, producing such exotic substitutes for photographic film as charge-coupled and charge-injection devices, offshoots of the semiconductor technology that has powered the computer revolution. These devices are capable of yielding pictures of high clarity and better geometric fidelity than can be obtained on photographic plates, crucial factors in the planet search.

These advances in electronics will be incorporated into a number of new instruments. First will come vastly improved astrometrical telescopes. The Barnard's star debacle does not diminish astrometry as a highly reliable technique; it merely points to the faults of one particular telescope and the way it was used.

In fact, the major emphasis in Project Orion has been placed on the design concept of a vastly improved ground-based astrometric telescope. It would yield a theoretical accuracy thirty to fifty times better than is currently obtained. David Black, the NASA astronomer at the Ames Research Center, notes that, with one exception, the telescopes presently used in astrometry are about sixty years old, and only two were designed for high-precision astrometric observations. The remaining telescopes had been converted to photographic astrometry use by the addition of a color filter; such a filter, it now appears, caused the spurious planet discovery with the Sproul telescope. What's more, none of the existing instruments were designed to meet the requirements of ultrahigh precision and long-term stability necessary for successful detection of very small perturbations.

So far, astrometric measurements have been good enough to detect stellar objects whose masses fall considerably below the theoretical nuclear-burning cutoff. But they haven't been good enough to show the existence of objects whose masses are less than ten "millisun," a millisun being one thousandth the mass of the sun, or equal to the mass of Jupiter.

This situation is now being rapidly corrected. Under construction at the University of Pittsburgh's Allegheny Observatory is a new type of astrometric photometer, scheduled to be completed by the end of 1981. This device could extend the detection of Jupiter-sized planets to at least forty light-years, and at the same time, because of its efficiency, drastically shorten the observing time from years to hours. Within that distance of the sun shine about 500 stars, some 130 of them sun-like, 330 red dwarfs, and the rest of other types. Lifted into space, the Allegheny Observatory's detector should be able to spot planets the size of the Earth orbiting any of the several hundred stars nearest the Earth. Preliminary tests of a prototype detector by astronomer George Gatewood and his colleagues early in 1980 hinted that the device might do even better than that. Near the closest stars, it may reveal planets only one tenth the size of the Earth. As a result, discovery of the first planets may come even as you read these words. And the same instrument could enable a detailed study of planetary motions light-years away, to reveal the characteristics of the planetary systems.

While astrometric techniques measure the gravitational effects of other bodies on stars, there is a related method that could also detect planets without seeing them directly. This involves the measurement of a star's "radial velocity," or its motion in our line of sight. The center of mass, or barycenter, of the sun-Jupiter system, for instance, is displaced a small distance from the center of the sun toward its outer layers, due to the presence of Jupiter. The two heavenly bodies orbit this common mass like two children on a teeter-totter, except for the sizes of their orbits. While Jupiter goes around the sun in a big circle once every twelve years, the sun, being the bigger "fellow," describes a twelve-year orbit in a tiny circle, rolling around on top of its surface somewhat like a mass of jelly in a bowl, at about forty feet, or thirteen meters, a second. This is the same speed at which Jupiter moves along *its* orbit around the sun; it is Jupiter that adds the forty-foot-a-second motion component to the sun's orbital speed. That motion is the sun's radial velocity, so called because it's measured along the sun's radius.

Theoretically, the effect of a Jupiter-sized planet on its star's radial velocity could be detected spectroscopically by recording the changes in the starlight as the star moves about its short orbit toward and away from us. Here the Doppler effect comes into play—the apparent change in the wavelength of light, or sound, caused by the motion of the source.

The classical example is of a whistling train passing an observer. As the train approaches, soundwaves enter an observer's ears more frequently, making the sound high-pitched. As the train passes by and recedes, soundwaves enter the observer's ears less frequently, and the pitch of the whistle drops.

So it is with starlight. As the star moves toward us, its light will be more energetic and will appear bluish. As the star recedes, its light will become less energetic and will appear reddish.

The starlight will alternate between red and blue with the same period that the unseen planet orbits. The star's radial velocity can thus reveal not only whether planets accompany it but also how big and how far away from the star they are, since each planet would impose its own radial-velocity component on the star, adding to the star's orbital speed in a cumulative way.

The trick is to detect this minute angular displacement of the star's spectrum. This method works for detection of stellar companions of parent stars, since these companions are much larger than planets and accordingly impose a greater orbital speed on their central star. For Jupiter-sized and smaller objects, however, this type of detection is beyond the capability of current instruments. But new ones are being developed, just as they are for astrometry.

The most advanced radial-velocity detector is under development at the University of Arizona's Lunar and Planetary Laboratory, in Tucson. This spectrometer is being perfected under the direction of Polish-born astronomer Krzysztof M. Serkowski, who has devoted twelve years to the instrument's evolution. Dr. Serkowski and his associates began preliminary observations with the new device late in 1980 and hope, by the end of 1982, to improve the spectrometer's performance to enable it to detect other planetary systems.

Even now, scientists feel, preliminary observations with improved radial-velocity detectors could lay the groundwork for the planetary search. Such observations could better characterize the frequency of binaries among solar-type stars and assess the magnitude and frequency of intrinsic stellar variations that could masquerade as planetary systems. The first kind of observation would clarify the question whether stars automatically form either companion stars or planetary systems or whether the evolutionary path of planetary systems is fundamentally different. The second kind of observations would show whether sun-like stars show periodic, or quasiperiodic, intrinsic variations in radial velocity of the type one would expect planets to impart to the stars. Obviously, no one wants a repetition of the Barnard's star fiasco.

In contrast to astrometric and radial-velocity studies, in which the presence of planets is inferred from their effects on the stars while the planets themselves remain unseen, much like the Cheshire cat in *Alice in Wonderland*, whose presence is inferred from the smile that marks the place where the cat had been, more direct methods will be employed too. The first technique will measure a planet's intrinsic heat through infrared interferometry, and the second will record the starlight reflected by the planets, photographing them directly.

Consider interferometry first. On the face of it, the idea of taking the temperature of a planet light-years away sounds improbable. It may be likened to taking the temperature of a particular locust in a swarm of locusts with instruments aboard a satellite orbiting the Earth.

Actually, the situation is much more hopeful than that. For instead of taking the temperature of a locust, scientists will be taking the temperature of, so to speak, a parasite of the locust. The locust in this analogy becomes the parent star and the parasite the planet. The parasite-planet has properties that will make it look different from the locust, its sun.

The reason for this hopeful picture is that when looked at in infrared, planets stand out much more sharply than in visible light, compared to their parent stars. While the sun shines two and a half billion times brighter than Jupiter in visible light, as we indicated earlier in another context, a planet the size of Jupiter when seen in infrared would be only ten thousand times dimmer than the sun—an almost incredible jump in visibility.

The interferometer will then systematically scrutinize such planets, measuring their temperatures. It may even be able to spot planets in the process of being born in their "nurseries" around young stars by detecting warm interstellar dust grains forming into larger bodies.

The indirect and direct techniques complement each other in terms of information they can gather. While both can yield data on a planet's orbital period, for instance, a planet's temperature, its diameter, and its atmospheric composition can be measured only directly. On the other hand, the planet's orbital inclination and its mass can be detected indirectly.

The two techniques are also complementary with regard to the types of planetary systems they are best suited to discover. Direct techniques have a better chance of locating bright, or hot, planets, which would be found around more massive, bright stars. Indirect techniques, on the other hand, are best suited for detecting planets that revolve around low-mass stars.

Both astrometrical telescopes and interferometers will do much better from space than from the ground, and space near Earth is where they will eventually go. Some such instruments may be mounted aboard the space telescope that NASA plans to send into orbit aboard the space shuttle in 1983.

Later still, toward the end of this decade, a special telescope will be built to take pictures of the faraway planets directly.

Because planets shine only by reflected light, they will be literally submerged in the glare of their primary stars. This will make detection of extrasolar planets by photographic techniques extremely difficult. Trying to photograph an extrasolar planet is somewhat akin to trying to photograph a firefly next to an automobile headlight from a distance of several miles.

A telescope's vision improves dramatically when it is placed above the distorting veil of the Earth's atmosphere. NASA's space telescope scheduled for launch in December 1983 will be the biggest telescope to be placed in orbit. A mirror 96 inches (2.4 meters) across, it should improve our ability to see the universe about tenfold.

About half the diameter of the great, 200-inch mirror atop Palomar Mountain, in California, the space telescope will allow us to see stars and other objects about fifty times fainter than are visible through the biggest Earth-based telescopes. The space telescope will be in a 300-mile-high orbit, accessible to the shuttle astronauts, who will be able to maintain, replace, and repair instruments aboard the free-floating telescope, which will be controlled from the ground.

Many astronomers are excited about the space telescope. Says Margaret Burbidge, formerly Britain's Astronomer Royal and now professor of physics and director of the Center for Astrophysics and Space Sciences at the University of California's San Diego campus: "It's going to be the biggest leap forward in the history of astronomy. It will be able to see the whole spectrum from the far ultraviolet to far infrared. It will be away from city lights, too! We have no idea what we'll discover."

But the first space telescope is not specifically designed to detect extrasolar planets. Scientists think it's highly unlikely that the first space telescope could photograph extrasolar planets, even if the telescope system were modified. But surprises are possible—the history of science is full of them. Margaret Burbidge, for one, thinks that extrasolar planets may show up on space-telescope images. NASA's David Black thinks that the space telescope may detect an unusual planetary system. But to conduct a systematic, deliberate search, he adds, a specially designed telescope is likely to be needed, one with a mirror so finely polished that it could collect light with a precision not attainable now.

Although very difficult, direct photography and infrared imaging of extrasolar planets would reveal some important facts that less-direct methods won't. Much could be learned about a planet's suitability for life through direct photography. The planet's color, for instance, if bluish in comparison with that of its parent star, would probably indicate the presence of oceans. Similarly, a greenish tint would suggest the presence of chlorophyll and green plants. Atmospheric components could also be determined from direct photography and other types of direct imaging. Methane and oxygen, for instance, are present in large quantities in the Earth's atmosphere, and from their presence alone it would be possible, using special detectors, to deduce whether life exists on that particular planet.

The big problem in direct photography of extrasolar planets is how to blot out the brilliant light of a star that has a retinue of planets, without losing reflected starlight from the planets. There are, however, a number of approaches to that, ranging from a possible blacking out of the telescope's "pupil" to the use of an interferometer in conjunction with the telescope. The interferometer gets its name from interfering with light; in this case it would interfere with the starlight but not with the reflected planetary light. An interferometer splits a beam of light in two. Since light from a planet, being reflected light, would have traveled a greater distance to arrive at the observer than the light coming directly from the star, the interferometer could easily separate the two.

In a 450-foot-long dark tunnel at the Ames Research Center, Project Orion scientists have tested a number of masks on a small telescope imitating the workings of a much larger one. Their model planetary system, situated at the other end of the tunnel, was a big box with a standard projection lamp as the light source and two pinholes to mimic a star-planet system. The pinholes were identical in size, and the intensity of the light transmitted through one of them, representing the planet, was controlled by neutral-density filters.

After successfully testing a model of the Sirius A- Sirius B (The Dog Star and its "pup" companion) system, the scientists then tested their model star-planet system. They found they could successfully resolve the "star" and its "planet" with a number of different masks. In the realistic-appearing photographs of the test, the "planet" shimmers as a small blob of light next to a spiked image of the "star."

An ideal mask for a space telescope designed to detect planets elsewhere, interestingly enough, would be the "black limb" of the moon —the portion of the moon not illuminated by the sun or sunlight reflected by the Earth.

The moon's limb would occult the star under study while letting starlight reflected by its planets reach the telescope, as shown in the drawing facing this page.

The near-Earth orbit would not be suitable for such a telescope, because the moon would not be sufficiently free of Earthshine. Furthermore, the relative velocity of the telescope and the lunar limb would be too great to allow the appropriate alignment to be maintained long enough. In addition, most of the nearby stars of interest would not be occulted by the moon if the telescope were in near-Earth orbit.

To overcome these difficulties, the space telescope would have to be placed in an orbit as high as the moon, and the telescope's position would have to be changed from time to time to achieve the proper alignment with the lunar limb.

From its high perch, a 96-inch space telescope would then be able to spot a planet the size of Jupiter orbiting any of the forty or so stars within thirty-two light-years of the Earth. If we were trying to detect Jupiter from that distance, it would appear about 30 million times dimmer than the sun, as seen through the telescope.

Once a faint image is detected near a star, several tests will be applied to confirm that it represents a planet. First, its intensity and distance from the star must lie within acceptable limits. Next, the proper motion of the suspected planet would have to be checked against that of the star, a task that may require several months of repeated measurements. If the motions match, that would be strong evidence that the body is indeed a planet. Depending on the clarity of the planetary image, it might even be possible to obtain a crude spectrum of the planet and search for the methane absorption bands in the far red—characteristic of Jupiter and the other large planets of our solar system. This same space telescope could also be employed for many other types of astronomical observations, to measure diameters of stars and other objects, for instance.

There are still other, less obvious techniques that will be tried to detect other planets.

Two of the methods involve detection of the passage of distant planets across the disks of their stars. In the first approach, scientists would try to cash in on the fact that a planet the size of Jupiter would block a measurable fraction of light from its star if it moved between the observer and its primary star, causing a partial eclipse of

the star. An observer watching the sun from a faraway planet could measure the amount of light lost, if his instruments were similar to ours. This event, though, would occur only once every twelve years and the observer would have to be lucky; he would have to be situated in the narrow section of space from where Jupiter's transit across the disk of the sun would be visible.

Twelve years is a long wait. But there are other ways of looking for planets crossing their stars, such as surveying a large number of stars of known luminosity at once in the hope of catching some planets in transit.

The second technique of this type would involve looking for color changes in a star as a planet crosses it. As a dark companion transits the disk of a distant star, it leaves a characteristic color signature. This signature appears as a slight shift toward the blue in the color of the limb of the star, followed by an abnormal reddening of the stellar disk, and finally by another shift toward the blue, as a planet or a stellar companion approaches the far limb.

These color changes probably could be detected photometrically. Some scientists believe that an observational system consisting of three telescopes at well-separated sites and capable of one or more planetary detections a year could be built at a reasonable cost.

Listening to natural radio noise from planets elsewhere may be still another way of detecting them. Such radio noise arises from the interaction of electrons in a planet's magnetosphere with its magnetic field. Such signals are quite intense, and in our solar system they have been associated with Jupiter, Saturn, the Earth, and possibly Uranus. Jupiter's signals are particularly pronounced, and the possibility exists of detecting such radio noise bursts from large planets elsewhere.

The distant planets may also give themselves away by emitting from their atmospheres natural radiation of the organized, maser-laser variety. (Maser stands for microwave amplification by stimulated emission of radiation, while the first letter in laser stands for "light." Masers and lasers emit microwaves and light, respectively, in an organized, or "coherent," manner.)

Charles Townes, one of the discoverers of the maser-laser principle, reported at one of the workshops on planetary detection sponsored by NASA that the atmospheres of both Venus and Mars seem to emit a narrow carbon dioxide spectral line of unexpected intensity, due perhaps to maser-laser action. The energy is pumped in by the sun, producing a narrow spike of excess radiation from the region of the planetary atmospheres exposed to the sun.

A planet may thus announce its presence with periodic flashes in microwave or visible light. Many radio astronomers think, in fact, that the moment of planetary formation could be detected, because a planetary system in its birth pangs would emit intense radio-spectral lines such as the water-vapor line—another example of anomalous maser-laser sources suggested by Townes.

Some thought is also being given to imaginative planet detectors that would operate far out in space. Shklovsky and another prominent Soviet proponent of the search for extraterrestrial civilizations, Nikolai S. Kardashev, along with a number of other Soviet scientists, have proposed the construction of an unusual radio telescope in space.

They visualize what they call the "infinitely built-up space radio-telescope," which would be up to six miles (ten kilometers) in diameter, assembled from many smaller modules, about six hundred feet (two hundred meters) across. Two or more such radio telescopes would be arrayed about twenty astronomical units apart. (An astronomical unit is the distance from the Earth to the sun, 93 million miles). One telescope would thus have to be placed in an orbit in the vicinity of Saturn. Being that far apart, the radio telescopes would act like two giant ears, able to produce three-dimensional pictures of the entire universe. They could determine the distance, size, and shape of every observable object in the universe, and detect infrared radiation of Earth-like planets one hundred light-years away, and that of Jupiter-sized planets one thousand light-years distant.

In addition, the world's largest optical telescope, four hundred inches across, is being designed by University of California scientists. It will not only provide a new look out to the edge of the universe but is also expected to help in the planet search.

To answer the question whether planetary systems exist around nearby stars, a modest effort may yield results. A larger-scale endeavor would have to be mounted to settle the more fundamental questions about the frequency of planetary systems and types of stars that possess such systems. Dr. Douglas G. Currie, a University of Maryland astronomer, and his colleagues have concluded in a study for NASA that new devices employing two systems, amplitude and speckle interferometry, could be used with existing ground-based telescopes to detect planets the size of Jupiter to thirty-two light-years out. The Currie group further estimated that a six-foot (two-meter) interferometric telescope designed specifically to monitor binary stars, and capable of detecting planets the size of Jupiter, could be built for $1.4 million—a pittance in terms of total federal spending. Even a larger-scale effort, including telescopes in space, is likely to cost no more than a typical spacecraft mission to one of the planets of the solar system, concluded Orion Project scientists.

What will be the best places to search?

Stars are divided into ten major categories, or spectral types, in the following manner:

TYPE	SURFACE TEMPERATURE		COLOR
	F	(C)	
W and O	64,800	(36,000)	Greenish white
B	50,400	(28,000)	Blue
A	18,000	(10,000)	White
F	13,500	(7,500)	Yellow
G (giant)	9,360	(5,200)	Yellow
G (dwarf-Sun)	10,000	(6,500)	Yellow
K (giant)	7,560	(4,200)	Orange
K (dwarf)	8,820	(4,900)	Orange
M (giant and dwarf)	6,100	(3,400)	Orange-red
R	4,100	(2,300)	Orange-red
N	4,680	(2,600)	Red
S	4,680	(2,600)	Red

Stars come in many sizes. Although it may slightly bruise our egos to hear the sun described as a yellow dwarf, it is actually a medium-sized star compared to much smaller and much larger counterparts.

Some stars are so huge that if placed at the center of the solar system they would subsume all, or most, of it. The stars Antares and Mira, for example, are 330 and 460 times, respectively, the sun's diameter. The giant among visible stars, Epsilon Aurigae, is three thousand times larger than the sun, with a diameter of 2.3 billion miles. Placed inside the solar system, Epsilon Aurigae's outer rim would reach as far as the orbit of Jupiter. Still larger stars, big enough to cover all of our solar system, are visible in the ultraviolet and X rays only, because of the nature of their structure and emissions.

It would be fun to visualize superplanets circling such superstars, planets of unimaginable size, where Gulliver would be turned into a submicroscopic Lilliputian. Planets that size may exist, but they are not likely to evolve life, because the giant stars don't live long enough. Generally, they are gone from the stable stellar "main sequence" (see Chapter One, "Building Life's Stage") in a few million years.

At the other extreme are the red dwarfs, which, as we have seen, account for 90 percent of all stars. Most of them are extremely small and dim compared to the sun. Even so, they would still offer a life zone of sufficient warmth to life-bearing planets, provided these planets are situated close enough to the dwarfs. In most cases, this would mean a location much closer than Mercury is to the sun since red dwarfs usually generate only half the sun's warmth.

Close proximity to the parent star could introduce a difficulty: the possibility of the planet becoming locked in its orbital and rotational motions due to gravitational attraction so that the same side of the planet would always face its sun. This almost happened to Mercury; it has a "day" 58.7 Earth days long and a "year" of only 88 days.

As a consequence, Mercury finds itself in an unusual kind of resonance with the sun: the planet rotates three times for every two orbits around the sun. This alternately exposes the surface of the planet to the sun's searing heat and then to a 29½-day night, when the temperature plunges to a numbing −279 degrees F (−173 degrees C).

Could life exist on a planet similarly situated near a red dwarf?

If it does, the unusual circumstances on the planet's surface would demand the evolution of an extremely adaptable type of life. The life-forms would be forced to migrate with the short seasons, to keep within the more habitable life zones.

Although an atmosphere thick enough may help moderate such unusual day-night cycles, as it does around our poles, this problem obviously reduces the attractiveness of planets locked in such unusual orbits as abodes of life.

Still another unknown is the inherent stability of the dwarfs themselves. Some are known to flare up rather violently, emitting ultraviolet and other radiation, which would bathe their planets. Exposure to such radiation bursts could hardly be encouraging to the development of life, although some scientists think that such radiation might speed up mutations. What's unresolved right now is whether the flares occur on all the red dwarfs or only on the younger ones. If the dwarfs settle down to a peaceful adulthood after a turbulent adolescence, the environment for life on their planets would become much more stable, of course, and more conducive to life.

While the red dwarfs aren't completely out, sun-like G and K stars obviously are better candidates. There are at least 20 billion such stars in our galaxy. Not all of them may have planets, of course. Struve's suggestion that G and K stars rotate slowly because of planets around them is now believed by most astronomers to be false; the slowness is explainable by causes existing within the stars. The current belief is that the stars rotate slowly because the solar wind, which consists of a plasma that continuously flows off the surfaces of those stars, has dissipated their rotation rate.

An indication of how many G and K stars do have planets comes from a recent survey of sixteen nearby stars of that type. All sixteen stars were found to be members of either binary or multiple systems. The sun would, then, be one of seventeen such stars to have planets.

Another survey of 123 sun-like stars, conducted by Dr. Helmut Abt at the Kitt Peak National Observatory, in Arizona, found that the average star in the sample had 1.4 detectable companions, most of them star-sized. Extrapolating from observed patterns to smaller masses than actually detected, Dr. Abt concluded that in short-period systems, where the outermost companion star rotates around the center of the system's mass in less than one hundred years, one third of the companions are likely to be smaller than stars. Abt guesses that about one tenth of the stars in his sample have planets.

In another study, Abt found that 16 percent of hot stars, with surface temperatures between 20,000 and 100,000 degrees Fahrenheit, have companions of subsolar mass, suggesting they may be planets.

The implications of these discoveries are staggering. The first implication is that stars resembling the sun have either stellar companions or planets. Obviously, the sample is too small to reach a firm conclusion on this point. But theoretical understanding of this question suggests that it is a valid picture for all sun-like stars. The second implication—that hot stars have planets—would raise tremendously the total number of planetary systems in our galaxy. In fact, on the basis of radial-velocity studies, astronomers now suspect that 50 to 75 percent of *all* stars are parts of either binary or multiple star systems, with companions with orbital periods of ten to one hundred years.

If this high ratio holds up throughout the universe, some binary

OPPOSITE: *Not one, but two or more suns may shine on some planets.*

and multiple systems would probably be disqualified as abodes for life-bearing planets because of the complexity and potential instability of the planetary orbits.

Not all binary or multiple star systems would be excluded, however. Stability of the planetary orbits would depend on the nature of the star systems. Within thirty-two light-years, for instance, there are 188 observable binary systems, in most of which, scientists believe, it would be possible to have stable planetary orbits similar to those found in our solar system.

Orbiting one of the stars in a binary multiple system, a planet would be bathed in the light of two or more stars at once, stars of differing sizes and differing colors, such as yellow, orange, and blue—an exciting setting for life, as shown in our illustration on the previous page.

Another place to look for planets may be the vicinity of white dwarfs, the shrunken, dying grandfather stars that remain after a star goes through a giant stage. During that phase, a star bloats into a gigantic sphere, knocking nearby planets out of their orbits and sending them into deadly spirals that take them to a fiery end in the furnace-like interiors of the giant stars. The Earth apparently has such a fate in store when the sun turns into a red giant, 5 billion years from now. But outer, Jupiter-sized planets are expected to survive the fiery holocaust and perhaps even grow slightly in size by accreting the substance of the stellar wind from the expanding giants.

Since typical white dwarfs are only about one to two ten-thousandths the size of the sun, big planets accompanying the dwarfs should stand out clearly when viewed in the far infrared by a space telescope, suggests University of Pennsylvania astronomer William M. Frawley.

Such big planets could be detected around the dozen or so white dwarfs known to be situated within thirty-two light-years of the sun. The planets should be discernible because planets that size generate large amounts of internal heat. Frawley suggests that an infrared detector used with the space telescope might quickly determine whether planetary systems are common.

It now appears that even pulsars may have planetary companions. A pulsar is a compact body, believed to be a neutron star. In its atmosphere it has a radio-emitting region created by the star's magnetic field. As the tiny star rotates—often with the incredible speed of a fraction of a second for a single rotation—its highly precise pulses cross our line of sight. The pulse rate of pulsars is generally slowing down, and some astronomers think that nearby planets are responsible. In the case of the pulsar known as PSRO329+54, a group of Polish astronomers using U.S. data recently concluded that the effect may be attributed to a small planet, only one fifteenth to one half the size of the Earth.

The ultimate, and the most satisfying, type of exploration of planets elsewhere is still in the future. One day, it should become possible to look for life in the vicinity of other stars the way we have looked for life on Mars and photographed the spectacular satellites of Jupiter from spacecraft flying nearby, detecting previously unseen and dramatic events like the volcanic eruptions of the Jovian satellite Io.

The bigger the planet the easier it will be to detect, of course. An alien starship entering our solar system and flying toward the sun, for instance, would easily spot the giant planets with their satellite systems: Neptune, Uranus, Saturn, and Jupiter. On the other hand, Mars, Earth, Venus, and Mercury would hardly be detectable until the spacecraft got in close.

Just how long it will take before we develop the capability to explore planets outside our solar system with instruments from flyby spacecraft, no one can say. Even the nearest star, Proxima Centauri, is very, very far away: 26 trillion miles. Light, at 186,000 miles (300,000 kilometers) *a second* takes 4.2 years to travel from Proxima Centauri to Earth. Our fastest spacecraft travel at about 18,000 miles *an hour*. Obviously, those speeds aren't going to get us to the stars within a reasonable time. In fact, to span the 4.2-light-year gulf of space to Proxima Centauri at 18,000 miles an hour, or five miles per second, would take our spacecraft 156,240 years.

In trying to figure out how to speed up the trip, a group of British engineers calculated that the travel time—one way—could be cut down to a more manageable fifty years or so by building a starship that would be propelled to nearly the speed of light by an unlikely sounding propulsion means: explosions of small hydrogen bombs. The theoretical spacecraft of their design, *Daedalus,* had Barnard's star as its destination because the possibility of its having planets then looked strong. *Daedalus* would have to gather its information very rapidly during the flyby, because slowing down the spaceship would put too severe a demand on its design.

That or a similar type of starship may not be built and launched for another century. But the realization that technology is either already in hand or soon will be developed for the great planet search is exciting enough. In some ways, it's hard to believe that within the brief span of our technological development we already stand on the threshold of detecting planets light-years away. As one participant in a recent NASA workshop on planetary detection aptly put it: "Finding other Earths, or even other solar systems, holds more potential excitement than quasars, black holes, and other such 'pop' science."

Unfortunately, federal support for this exciting search is far below a desirable level; the United States may be the richest country on Earth but in today's austerity climate, basic science is being cut back—unthinkingly and at the expense of the future. To help keep the planetary search going, George Gatewood and Frank Drake recently formed the Extrasolar Planetary Foundation, with headquarters at the Observatory Station, Pittsburgh, Pennsylvania 15214. Writing in *Cosmic Search*, a magazine published on a shoestring by John Kraus, director of the Ohio State University radio observatory, Gatewood explained the Foundation's aim. "With a board of directors composed of astronomers and lay persons with related backgrounds, we hope to act as a focal point for those who would like to give this new effort a measure of support," he wrote. "Our experience suggests that the number of people who would like to participate in this effort is large enough to raise an endowment of several hundred thousand dollars. The interest earned by such a fund would become a perpetual source of support, providing new instrumentation and providing the long-term stability necessary for the observational programs that will be required."

As for the impact of such a discovery, if the first pictures of the fragile and beautiful planet Earth taken from space had an immediate and profound effect on man's thinking, the discovery of planets elsewhere is bound to have an even more dramatic influence. The value of cosmology, which deals with the large-scale structure of the universe, in helping man comprehend the grandeur of the universe and his place in it, has long been appreciated. Now cosmogony, the other extreme of astronomy, which studies the formation of planetary systems, promises to help us ultimately to discover other forms of life, thus opening new perspectives on the definition of life and on the meaning of man's existence in the universe. Not only will the impact of the discovery of extrasolar planets on human philosophy be possibly greater than that of any other scientific advance of the past, but the challenge to contact and even visit those newly discovered worlds could well become the driving force of space exploration for decades, even centuries to come. We will discuss in the next chapter how we may be able to establish contact with the inhabitants of those worlds.

CHAPTER TEN

Contact!

Any day now, if radio astronomers are to be believed, a signal from the stars will flash across the unimaginable gulf of space to end our cosmic loneliness. The signal may sound like a piano concerto or the organ-like tones of the theme song of the Hollywood movie *Close Encounters of the Third Kind;* those call letters were fashioned after the sounds of a coded message beamed from Earth toward the stars in 1974. Depending on the type of information encoded, the sounds from the stars may not be as pretty as music. They may barge in instead like the buzz of an angry bumblebee, not implying that it was an angry message but signifying a television-like transmission. The musical tones, on the other hand, would tell the scientists that a pictorial coded message is coming in.

That the signal will come one day soon, there is no doubt among radio astronomers. For many years it looked as if the search for the faint whispers in the abyss of space would prove an almost hopeless task that would take centuries, perhaps even millennia. But now the odds have taken a decided turn for the better. Vastly improved detecting apparatus is being built; it may suddenly reveal signals that we have been unable to detect before, think the signal searchers.

With their new, highly sensitive "ears," at any rate, radio astronomers feel they should be able to listen to all the billions of radio frequencies, covering them in decades instead of centuries.

The radio astronomers may be badly mistaken, of course, in their almost religious belief that other civilizations have developed along technical lines similar to ours. Instead of being another radio astronomer, an intelligent creature elsewhere, as we have seen throughout this book, is much more likely to be an intelligent descendant of an octopus, of an insect, or of a reptile, possibly without any interest in, or aptitude for, high technology, space travel, or interstellar communication.

It appears from what we know about evolution of life on Earth and from the recent detection of life's building blocks among the stars, that we have a lot of cosmic cousins—some of them rather strange-looking characters, as the illustrations in this book show.

But we are unlikely to have cosmic brothers and sisters. Not only the basic nonrepeatability of evolution argues against exact duplicates of humans existing anywhere else. The timing, the coincidence of two such sets of beings existing at approximately the same time strains credulity. In the 3.5-billion-year history of life, man has been on stage for a fleeting instant, for about one hundred thousand years in his present form, a mere thirty-five-thousandth of the time—a second in the centuries of life's span. And radio in its present state is only a few decades old. For two or more societies to be at exactly the same point in technological development seems highly improbable.

With most stars being older than the sun, if technological civilizations exist, they are likely to be unimaginably more advanced than ours. They could have relegated radio to their attics long ago. Even one of the most vociferous proponents of the search for extraterrestrial intelligence by means of radio, the Cornell University astrophysicist Carl Sagan, has conceded that much. At a conference on the subject, Sagan admitted that other civilizations' communications techniques may be so advanced as to be largely inaccessible to us. "It may be," Sagan added, "that we are very much like the inhabitants of, let's say, isolated valleys in New Guinea, who communicate with their neighbors by runner and drum, and who are completely unaware of a vast international radio and cable traffic over them, around them, and through them."

So if the central idea of this book—the uniqueness of man in a Darwinian universe—is correct, why waste time and money on trying to search for extraterrestrial intelligence by listening to or sending signals into space?

The search for radio signals must go on because it's the only type of electromagnetic signal we are best equipped to detect and send over astronomical distances. In using radio, we may be like a man looking for a key that he has lost on a dark street with a few lights. The man will look for the key where there is light, not where it is dark, because that's all he can do at the moment. Still, considering the vastness of the universe, who can really say that a few civilizations might not be still using radio, to contact emerging technological societies such as ours? Billions of planetary systems are, after all, believed to be in our galaxy alone. And another 100 *billion* galaxies shine in the observable universe. To believe that life on Earth is an exceptional phenomenon would be an exceptional act of faith.

The problem of communication with other life-forms, to be sure, is a much more difficult task technically than the search for planets outside our solar system, described in the preceding chapter. The planet search is basically a scientific undertaking, with solid knowledge and known instrumentation backing it. The search for signals from the stars is a shot in the darkness of the cosmos, based largely on guesses and speculation, albeit highly interesting ones.

The search must go on because if we didn't try we would never know whether intelligent whispers even now rustle through the spaces along the stars.

The recognition that extraterrestrial civilizations might exist and might be sending out messages is not new in myth and folklore but is new where modern science is concerned. As the capabilities of radar and radio astronomy grew rapidly after World War II, it was only natural for enterprising young scientists to start thinking seriously of trying to contact other civilizations, first by listening to any signals they might be sending and then by beaming messages at the most likely stars.

Frank Drake is the pioneer of what has become known as the search for extraterrestrial intelligence, or SETI for short. Even as a boy growing up in Chicago, Frank Drake dreamed of other worlds. Later, he got a Ph.D. in astronomy from Harvard and in 1958 became a radio astronomer at the National Radio Astronomy Observatory, near remote Green Bank, West Virginia.

Drake was fortunate to be associated with Otto Struve, then the director at Green Bank and one of the few prominent astronomers who was not afraid to voice his belief in a populated universe. Struve, as we have noted in the preceding chapter, had determined that sun-like stars rotate slowly and ascribed that fact to the presence of planetary systems slowing down the spin of their suns. Not surprisingly, Struve egged Drake on to hurry with his idea to listen to the stars. Drake named his effort Project Ozma, after the Queen of Oz, the land of fancy created by L. Frank Baum.

In the midst of the preparations for Drake's momentous project, in 1959 Philip Morrison and Giuseppe Cocconi, at that time both professors at Cornell University, in Ithaca, New York, published a short paper in the British journal *Nature*, suggesting that a search for extraterrestrial signals might be worthwhile along the 21-centimeter hydrogen line in the electromagnetic spectrum, where the motion of interstellar hydrogen atoms generates radio noise.

This lent prestige to Drake's proposed search, but it also made it look, wrongly, that he was acting on Morrison's and Cocconi's suggestion. Drake had arrived at the idea independently. He had chosen the 21-centimeter frequency for somewhat more practical and prosaic reasons, however. Equipment so constructed could be used for conventional radio astronomy, so no one could accuse the scientists of wasting money. (It turned out that the whole project cost only two thousand dollars.)

Morrison's and Cocconi's rationale for the use of the 21-centimeter frequency, Drake has said, was perhaps more controversial, but far more stimulating, than his. The two physicists had suggested that the 21-centimeter line, being the emission line of the most abundant element in the universe, could be a communications link uniting a network of galactic civilizations. From that idea arose the "waterhole" concept, the idea that the 21-centimeter frequency could also be a beacon to attract emerging civilizations and bring them into the galactic club—a waterhole, in other words, where civilizations meet.

Finally, in the early-morning hours of April 8, 1960, Drake and his associates pointed the observatory's 85-foot dish at their target, the first of two relatively nearby stars. The first star, Tau Ceti, yielded no signals, but from the direction of the second, Epsilon Eridani, a strong pulsed signal began to boom as soon as the telescope had been pointed at it. The breathless listeners heard bursts of noise coming

out of the loudspeaker eight times a second. They moved the
receiver off the star and then back on it to make certain it was indeed
coming from the star, but in the brief interval the signal had disappeared. A week of listening failed to locate it again. About ten days
later, though, the signal came back, but, with a smaller, auxiliary
radio telescope pointed out the observatory window, Drake could
tell that the signal came from a high-flying airplane. This was the
first "birdie," the radio astronomers' term for man-made or natural
signals masquerading as signals from the stars.

Drake was not surprised by his failure to detect any signals, because he and other radio astronomers estimated that only one star in a
million may have a signal-generating planet.

After two months, Drake terminated Project Ozma, but the search
was soon picked up by scientists in the United States, the Soviet
Union, Canada, Great Britain, and West Germany. These searches,
however, have been sporadic and short-term. Of the 20 million stars
in our galaxy that are considered to have life-bearing planets, a mere
one thousand have been searched so far. The only American full-
time effort has been going on since 1973, but on a shoestring basis.
At Ohio State University, astronomers John Kraus and Robert
Dixon, and their students, have been engaged in a continuous search
with an antenna twenty-four thousand square feet in size. Their electronics, however, are obsolete, and the project is being conducted
without support of any federal agency.

Urging on the listeners has been their belief that technological civilizations are common. Frank Drake thought up the original formulation for the number of technical civilizations in the Milky Way galaxy at or beyond our level of technological advance. In this formula,
N, the number of such civilizations believed to be in existence in the
galaxy, is written as the product of seven factors:

$$N = R_* \, f_p \, n_e \, f_l \, f_i \, f_c \, L$$

R_* is the rate of star formation averaged over the lifetime of the
galaxy, in units of number of stars per year.

f_p is the fraction of stars with planetary systems.

n_e is the mean number of planets ecologically suitable for life.

f_l is the fraction of planets where life actually exists.

f_i is the fraction of such planets with intelligence in some form.

f_c is the fraction of planets on which intelligent beings develop an ability to engage in interstellar communication.

L is the mean lifetime of a civilization in a communicative phase.

The reliability of the formula declines sharply from R∗ to L, and the equation, of course, involves studies uniting diverse disciplines, from astrophysics to biology, psychology, sociology, and politics, all of varying veracity.

The formula starts off with a statistical quantity, the rate of star formation, which is determined by actual counting and some extrapolation. But, going on, scientists have to extrapolate from fewer and fewer examples—in the end, from a single one: that of Earth and our kind of life. Finally, when they reach L, the mean lifetime of technical civilizations, there are no examples to go on, since, fortunately for us, our civilization still exists. Still, as one scientist wrote recently, "the concept of subjective probability is at present the only basis upon which probability estimates can be made about extraterrestrial intelligent life."

The lifetime of civilizations, of course, would be the crucial determining factor in how many civilizations might be around for us to communicate with. If civilizations destroy themselves shortly after developing radio technology, then there would be only a handful of them in the galaxy, and the chances of finding them would be pretty slim. If, on the other hand, civilizations persist for billions of years, then there would be billions of them around—another unlikely thought.

The radio astronomers hope that sapient beings on other planets might have converged with us in behavior, much as, say, the extinct reptile ichthyosaur resembled the mammal dolphin. "They might have evolved to culture, and then, say, to radio telescopes!" a recent NASA report states hopefully. "Culture," it adds, "is a workable way of life, like hunting schools of mackerel."

But not all civilizations need be technological. We have seen this on Earth. The Mayas of Mexico and Central America boasted excellent astronomers but had no wheeled vehicles, no draft animals, and only simple forms of agriculture. The Mayas *did* possess writing and highly developed social and political organizations that marked other early civilizations, indicating that a civilization can flourish without higher technology.

We know, furthermore, that men as intelligent as we are live in societies without technology even today. The Kung bushmen of southern Africa, typical of other "Stone Age" people still living, exhibit an intelligence entirely comparable to that of "civilized" men, with impressive communication capabilities, yet they practice no agriculture and have only dogs as domesticated animals, suggesting the coexistence of a highly developed intelligence with a technical simplicity.

This could be the main reason why all that the signal searchers have heard so far is the gentle hiss of cosmic radio static—and those ubiquitous birdies.

The natural radio emissions from the galaxy sound much like the static heard in a radio receiver during a thunderstorm, intermingled with a hissing noise. Among the sources of natural radio noise in space are electrons spiraling in the magnetic field of our galaxy, powerful quasars and pulsars, as well as stars and clouds of gas and dust far out in space. There is also leftover radiation from the primordial fireball, the big bang that gave birth to the universe. To complicate matters, a noise similar to the galactic emissions is manufactured by the random motion of molecules within the receiving antenna itself. And the Earth's atmosphere generates radio noise too.

One way to determine whether the radio signal is originating in deep space and not in the Earth's atmosphere or in the antenna itself is to switch the antenna back and forth between the source of the signal and the nearby sky. Spurious noise from the antenna or from the atmosphere will be the same regardless where the antenna is pointed, but pointing the antenna back at the signal source will indicate that the signal is apparently coming from deep space if the signal remains stationary in the star field. If it moves, it is obviously an airplane or a satellite.

Most birdies can be dismissed for what they are this way. On occasion, however, particularly mysterious birdies can raise questions in the minds of even experienced radio astronomers.

A few years ago, for instance, Frank Drake, who is now director of the National Astronomy and Ionosphere Center, which operates the world's largest radio telescope, the 1,000-foot dish at Arecibo, Puerto Rico, tracked a tantalizing signal for hours that seemed to originate in the field of stars. But some hours later the signal faded out, and later still it appeared that it might have been coming from a weather balloon that had temporarily become stationary high up in the atmosphere, its drift rate matching the hardly perceptible motion of the stars.

Soviet scientists, on the other hand, have been misled on occasion by transmissions from secret American spy satellites into thinking that they were receiving signals from another civilization in space. They have even hinted to the press that they might have received an extraterrestrial signal.

Perhaps the most exciting birdie of all, however, involved the recording in 1967 of radio pulses that were being repeated with exceptional precision, a few times a second, and came unmistakably from the stars. Astronomers at Cambridge, England, kept quiet about the detection for a while and had even tentatively labeled their signals as coming from "Little Green Men." But a closer look at the pulses revealed an ever-so-slight slowdown in their timing, which clearly meant that the pulses were of natural origin. And so it turned out to be, for this was the discovery of the first pulsars, rapidly spinning neutron stars that create their own radio signals by rotating their magnetic fields.

Aside from the sometimes irritating birdies, another, much more serious problem has confronted the signal searchers in the past few years. Gradually, it became clear that the 21-centimeter hydrogen line wasn't all that unique. Many other frequencies, such as those of water vapor and of the hydroxyl radical (a fraction of the hydrogen molecule) were found to be almost as universal as the hydrogen line. In all, it appeared that billions of frequencies would have to be checked.

This situation obviously complicated the search strategy. With this diversity of possibilities, there was a fear that the search would consume a hopelessly long time, perhaps centuries. "Although we have the power to discover civilizations," a downcast Frank Drake told a symposium on SETI in 1973, "we know neither where to look nor on what frequency."

But, toward the end of 1980, a dramatic development so improved the SETI chances that a happy Frank Drake could now say: "It's no longer a question of centuries or millennia but of decades."

What had happened was that NASA finally got a small appropriation from Congress to start SETI as an official program. The question of what frequencies to choose for the search still remains unresolved, but now instruments are being built that will automatically listen to millions of channels at once, reducing the time needed for a large survey to more manageable proportions.

The SETI instruments will automatically sift through the signals and record them; a bell may ring to alert nearby scientists if an interesting signal is received. "The vast majority of the data that will be coming into this machine will be thrown away," explains Robert Edelson, until recently SETI project director at NASA's Jet Propulsion Laboratory, in Pasadena, California. "There will be so much data that humans could not possibly look at it all. So you have to very carefully define what is interesting so the machine knows what to keep for you."

The SETI instrumentation consists of an extremely large radio-spectrum analyzer and a signal-extraction device. The spectrum analyzer is the easier one to build; the signal-extraction device is much more demanding. The signal detector will be automatically tuned, somewhat like a home radio receiver, except that it will be able to receive anywhere from one to 10 million radio channels at once, up dramatically from the mere three thousand that it has been possible to survey at once until now, although on occasion the Arecibo telescope has been able to look at as many as sixty thousand channels at once. Improved automatic identification of birdies is one aspect of signal detection the engineers at Jet Propulsion Laboratory are working on.

Existing radio telescopes at Arecibo, Green Bank, Ohio State University, and various NASA tracking stations around the world will be equipped with copies of the sensitive SETI receivers and multimillion-channel analyzers. High-sensitivity searches will then be mounted along the "waterhole" frequency range, concentrating on nearby stars, star clusters, galaxies, and the plane of the galaxy. In addition, the entire sky will be searched with lower sensitivity over a wider frequency range. As the SETI leaders like to point out, the cost of the whole program, $30–40 million, would be comparable to the production costs of *Close Encounters of the Third Kind*.

So sensitive will be the new SETI signal-detection equipment that a signal from a taxicab transmitter on Earth reflected off an orbiting satellite's outer shell could saturate several channels of the receiver, producing an enormous power reading. "The system will have to be extremely sensitive," says Bob Edelson, "but because of that sensitivity, man-made signals are a real problem. And in fact, the reason we feel some urgency in getting this project rolling is because people are using more and more of the radio spectrum and it's getting more difficult to conduct observations looking for extraterrestrial signals."

As for the best frequencies to be probed, the region between the wavelengths of 1 millimeter and 30 centimeters, or 300 to 100 gigahertz, constitutes a relatively quiet reception area where both galactic and atmospheric noise are at their minimum. Within this window lie the 21-centimeter hydrogen line and the 18-centimeter hydroxyl line. The famous waterhole is between these two frequencies.

A number of other possible frequencies have been suggested. The water molecule, for instance, emits at both 2 millimeters and 13.5 millimeters. At 1.7 millimeters the sky background radiation from the primordial fireball is at its peak; the Soviet astronomer Nikolai Kardashev has suggested this frequency as another possible signpost. Still another frequency, that of 11.7 centimeters, has been cited as reflecting universal fundamental constants of nature.

Bob Edelson and other radio astronomers are convinced that radio is the best means of communicating across interstellar distances because, of all known particles, photons that comprise a radio beam are the easiest to generate in large numbers, and the easiest to focus and capture. Neutrinos, the invisible particles that appear to have little or no mass and travel at the speed of light, would be difficult to generate and detect as communications carriers. So would gravitational waves, electron beams, and other suggested substitutes for radio. "Unless there is a physics that we don't know at all," says Edelson, "electromagnetic waves are the best means of interstellar communication."

Furthermore, radio astronomers like to stress that radio technology today is almost mature—a great leap since that memorable day in 1901 when Guglielmo Marconi sent the first radio message across the Atlantic, a coded letter "S." Radio, claim the specialists, is near perfection technically. Edelson says, for instance, that there are fundamental limits to the improvement of a radio transmission by means of modulation and coding. "We can't identify a scheme that will get us to the limit," he says, "but we know the direction we have to take to get there. We are presently pretty close to that limit." He marvels that in the short time radio technology has been around, radio has almost been perfected.

Ironically, this very efficiency of our radio engineers complicates the signal search. As the signal is made more efficient, to be sent over interstellar distances with the least power, perversely it begins to look and sound more and more like the natural background noise. This happens because the band of frequencies that the signal must occupy gets wider and wider; this dispersion is what makes the signal look more like noise or static than a clear-cut signal.

In fact, to a radio ham in Borneo, signals from the recent American Jupiter and Saturn probes, for instance, would have already sounded like meaningless static. But when NASA's big radio telescopes receive those signals, they process them into a stream of information. The catch to successful deciphering is the need to know how the signals were modulated, or encoded, for transmission. "But if you don't happen to know how it was modulated," says Edelson, "you don't even know the signal is there, because it looks so much like noise that it blends into the natural background. And since there is literally an infinite number of ways to encode that signal, you can't reproduce the decoding method from any first principles."

For those reasons, Edelson thinks that eavesdropping on other civilizations' radio and television transmissions would be extremely difficult. He suggests that if a signal from the stars is detected, it's much more likely to be a "primary" signal that is intended to be intercepted because another society is eager to find its cosmic companions. This type of signal presumably will have some readily recognizable characteristics, such as rapid pulsing, to make it stand out as an artificial signal.

Radio astronomers are hopeful that civilizations in the cosmos will be generating such beacon signals to simplify our search. The beacon may not even be a radio signal. One unusual way to announce a civilization's presence, suggests Frank Drake, might be by inserting an artificial spectral line in the light of a star. A mere one hundred tons of suitable material that would emit a detectable artificial spectral line, placed in orbit around the sun, would be sufficient, says Drake.

Another astronomer, Neville Woolf, of the University of Arizona's Steward Observatory, has suggested—and investigated—another kind of optical signal, one that would be visible to the unaided eye at a distance of six to ten light-years, the distance of stars nearest to us and visible from the northern hemisphere. Woolf speculates that a civilization elsewhere might use comets as signal flares by igniting them with atom bombs, with the intent of alerting other civilizations that the senders exist. "The observing civilization might return the signal," says Woolf, "and the two could then engage in direct radio communication."

A comet would be an obvious choice, because it can be taken apart readily, Woolf argues. If we or another civilization could disrupt up to six comets a year, Woolf adds, the senders might be able to spell out the number pi, the ratio of the circumference of a circle to its diameter, in binary code.

Igniting cometary dust and slush for interstellar signaling may sound like a modern-day update of the nineteenth-century proposal to signal the Martians by digging a twenty-mile ditch in the Sahara, filling it with gasoline, and setting it on fire. But no idea, however unlikely at first glance, can be disregarded, especially if it's easy to check out. Woolf did check it out by looking at a few sun-like stars. So far, he hasn't seen any cometary bonfires blazing in the cosmos.

Detecting radio signals, on the other hand, will be like looking for "the thin gnat voices" that "cry star to faint star across the sky" in the words of Rupert Brooke in *The Jolly Company*. First, the interstellar distances will reduce the strength of even the most powerful radio signal. The strength of a signal ebbs in intensity by a factor of four each time the distance from the source doubles. Second, as we have seen, if rapid progress in our radio technology is any guide to what to expect, fishing a signal out of the natural background noise will be extremely difficult.

Even if a signal is received, there is no guarantee that it will be understood or even deciphered. Philip Morrison has noted that it took contemporary scientists ten years to fathom what Einstein meant by his 1905 formulation $E=mc^2$ (where energy equals mass times the speed of light squared), and the formulation was much more in the context of existing knowledge than any extraterrestrial message can be expected to be. In fact, some scientists who are not radio astronomers think that the probability of mutual intellegibility between interstellar searchers is very small. Dolphins illustrate the problems involved. We are still far from deciphering their complicated language of clicks and whistles or from gaining an insight into their ways of thinking and the real extent of their intelligence.

Radio astronomers and other physical scientists, on the other hand, are convinced that a common language—mathematics—will unite interstellar intelligences. It lends itself to being put into a form that is easily transmitted, they point out. By extension, another intelligence will know and understand physics, too, another common meeting ground, add these scientists. And on top of that, messages most likely to be received will be decodable into pictures, even possibly three-dimensional movies, they hope. The SETI proponents feel, in other words, that those who think that it will be difficult, if not impossible, to engage in a dialogue with another civilization in space are underestimating the amount and type of knowledge we would automatically have in common with other intelligent beings. The message, Morrison has suggested, because it would have been planned with extraordinary care, would not be as easy to read as a newspaper story but would be analogous instead to "a rich, difficult textbook on an advanced subject, full of diagrams, hints, and examples." (Quoted in *Communication with Extraterrestrial Intelligence*, edited by Carl Sagan, MIT Press, 1973.)

Interstellar distances would obviously put a crimp into any idea of telephone-like chit-chat with other civilizations. While it will be less than a ten-year round trip for a message to the nearest star, the distance from one end of our Milky Way galaxy to another stretches to 90,000 light-years. Exchanging a message with a civilization at the other end of our galaxy would take 180,000 years. Communication with another galaxy would take millions of years. Even Andromeda, the nearest galaxy, is almost 2 million light-years away, while more distant galaxies are billions of light-years distant.

Obviously, it would be more interesting to learn of another civilization five to ten light-years away instead of one thousand so that reasonable two-way contact could be established. Receiving a message across a 1,000-light-year distance would be analogous to receiving messages from the ancient Greeks—fascinating and important in itself but without a chance for a real two-way conversation.

If most people dislike the one-and-a-half-second delay inherent in the transoceanic telephone calls retransmitted through an orbiting communications satellite, to the point of suspecting the motives of their conversation partners, who seem to pause before replying, what dark suspicions could be entertained about interstellar conversationalists who delay their answers for centuries?

By the time they get here, furthermore, the signals could be artifacts of a society that has vanished long ago, "electromagnetic relics from antiquity rushing through space at the speed of light," as John Kraus has put it.

There is an almost child-like belief among some radio astronomers about the vast benefits those advanced beings could pass on to us: the secrets to the survival of an intelligent species, suggestions on how to avoid thermonuclear war, how to defeat cancer, and similar fantastic gifts. Such beliefs reflect man's egocentric nature and disregard the fact that intelligent beings elsewhere most likely are not man-like and don't necessarily engage in man-like foibles and absurdities or are subject to exactly the same diseases.

And there is an almost equally child-like fear of contact, among some other scientists. No less an authority than Sir Martin Ryle, a Nobel Prize-winning British astronomer, has urged SETI participants to listen only and not to send any signals into space. Although radio contact would be a meeting of the minds without physical encounter, Sir Martin, along with some other people, fears even radio contact, lest it bring disaster to Earth in the form of aliens eager to exploit humans, or even literally devour them. More likely than not, such fears are groundless because any civilization capable of space flight should be able to create whatever resources it wants almost at will and·would be unlikely to covet our rapidly diminishing natural riches.

The consequences of radio contact between two societies at different levels of development would be much unlike direct confrontation, of course. Such contact would not threaten us militarily or economically, as has often happened when a superior society has met a less-developed one face to face. But there is obviously the possibility of a large psychological shift, perhaps of the type that changed the course of the Japanese society when a mere threat of invasion of the new Western ways ended the centuries-long isolation of Japan by opening that country to an influx of new ideas and to people from other lands. With that psychological shift in Japan, eventually came a tremendous gain in scientific and technological capabilities that startled the West; so it may be with information gained from another civilization in space. The discovery of another civilization could thus positively influence the future of mankind.

It has been suggested that even our first contact with an extraterrestrial civilization will put us in touch with more, possibly many more than one such civilization. It may well be that one communicating civilization influences others to become such societies as well.

The idea that we should keep silent and not announce our presence with radio messages to the stars is simply behind the times. Willy-nilly, ever since Marconi first beamed his pioneering message across the Atlantic, the Earth has been rapidly turning into an electromagnetic beacon, as not only radio but also television and radar pulses have begun to leak into space. All this cacophony of sounds is spreading like a spherical shell from the Earth, much like a circular wave produced by a pebble dropped into a quiet pond. Thanks to the growing variety and strength of these signals, the Earth is now the brightest-shining radio beacon in our galactic neighborhood.

Some people have speculated that such old TV hits as "I Love Lucy" are now entertaining creatures on other planets. This is unlikely, because the portion of the carrier wave that carries the programs is not strong enough to be deciphered as a program at interstellar distances, although the signals themselves, now coming from the world's fifteen thousand TV stations, will appear unmistakably artificial. None of these signals, furthermore, has yet gone beyond about fifty light-years—an insignificant distance in cosmic terms. We may have announced our presence, but our "shout" is yet to echo through many stellar neighborhoods.

In addition to radio and TV signals streaming outward from the Earth, another type of intelligent communications activity—if television can be termed intelligent—should one day attract observant life-forms elsewhere. These are the radar-like beams of the great radio telescopes. These telescopes function in two modes: They are capable of listening and of sending, too; their radar beams are used to probe surfaces of the planets in our solar system, for instance. The beam from the biggest, the giant dish at Arecibo, as long as it remained pointed at the distant observer, would appear at its radio frequency as a star a million times brighter than the sun even thousands of light-years away. There are also about twenty extremely powerful military radars around the world, whose beams should be detectable at interstellar distances. Someone is bound to wonder what all those electromagnetic flashes are. "It's tantalizing to realize," concludes a recent NASA report on SETI, "that if another intelligent species should somehow recognize the solar system as a likely site for intelligent life, then it would be trivial [for that intelligent species] to illu-

minate it with an easily detectable signal from enormous distance."
Tantalizing it is, but we should keep in mind that radio telescopes
and big military radars are even younger than radio and television, so
their transmissions have traveled only a few light-years, just far
enough to reach only the nearest stars.

If there are any civilizations with an ability to communicate with
us within thirty light-years—which seems unlikely—we would not get
a response from them, at best, until the end of this century. And if
such civilizations are hundreds or thousands of light-years away, we
would not hear from them for centuries or millennia to come.

However, we might detect other civilizations even if they beam no
specific announcements toward us. If, as some scientists believe, a
wealth of astronomical and cultural information would be available
to a distant observer carefully monitoring the frequency and intensity variations of electromagnetic radiation leaking from Earth,
wouldn't it be natural for us to try to eavesdrop on other civilizations' transmissions?

The idea is attractive but even more difficult to execute than the
detection of deliberately sent interstellar signals. "Local" transmissions would be all that much weaker and more difficult to detect
from Earth. That's why a group of American scientists a few years
ago proposed the construction of an immense "orchard" of one thousand radio telescope dishes each three hundred feet in diameter; a
portion of this structure is shown in our illustration.

This arrangement, shown in the illustration on the next page, is
known as the Cyclops system. It was proposed by Dr. Bernard
Oliver, a SETI enthusiast who recently retired as vice-president for
research and development at Hewlett-Packard Corp., in Palo Alto,
California. The system of steerable radio telescopes would form, in
effect, one huge antenna covering twelve square miles (thirty square
kilometers), or more, all under computer control. Such a system
would take ten to twenty years to construct, but it could begin
searching nearby stars with only a few antennas operating and then
carry the search farther and farther into space as more antennas are
added to the array. The antenna farm could scrutinize, one at a time,
one million stars within one thousand light-years of the Earth. It
could not only search for radio beacons but eavesdrop as well.

Advanced civilizations could manifest their presence in other ways
besides radio signals. Ronald Bracewell, a Stanford University radio

astronomer, notes that in addition to emitting electromagnetic radiation, the Earth has become a modest source of material objects ejected into space, of infrared radiation resulting from fuel consumption, and perhaps of strange chemicals. A civilization much more advanced than ours, Dr. Bracewell notes, could be a strong source of any of the above emissions, and in addition might generate X rays, gamma rays, magnetohydrodynamic waves, high-energy particles, and unusual nuclides, or atoms with distinctive nuclear structure. Some of these emissions do not travel fast and do not necessarily arrive from the direction of the source, even though the manifestation might be spectacular, Bracewell suggests.

Extremely advanced civilizations could be engaged in cosmic engineering on an interstellar scale that would be conspicuous at distances of thirty-five hundred light-years and still detectable at even greater distances. No one knows, of course, if such civilizations exist, but the idea has become established in the SETI literature. The supposed structures of such civilizations are known variously after their creators, as the "Dyson Spheres" (after Freeman Dyson, a Princeton University physicist), or Kardashev Type II and Type III civilizations (after the Soviet astronomer Nikolai Kardashev).

In Dyson's scheme, a highly advanced civilization takes some of its solar system's larger planets apart and, using this matter, builds a sphere around their sun to utilize its full energy output. The Earth receives only one three-hundred-millionth of the total energy emitted by the sun in all directions at once. In contrast, Dyson's creatures, by settling on the interior of their spheres like flies on the inside of a lampshade, bask cozily in the total warmth of their sun.

Kardashev, for his part, has classified civilizations into Types I, II, and III. Type I is similar to the Earth in terms of its power output for interstellar communication. Type II can generate the power of the sun, and Type III that of a galaxy. Respectively, these civilizations are capable of restructuring planets, stars, and galaxies. Their activities, Kardashev has proposed, can be detected from afar.

Kardashev's colleague Iosif Shklovsky has speculated that Type II and Type III civilizations would not be biological but "computer-devised and spread over enormous areas." He thinks that existence of biological systems in environments that command energy resources 10 billion times greater than the energy at our disposal would be difficult, if not impossible, because of the great amounts of potentially lethal radiation that would be generated. Such speculation obviously crosses into science fiction, but we have to keep an open mind about the possibilities, since, a century from now, or even less, we may be able to build machines with some signs of intelligence.

If nothing else, the acceptance of such ideas by conventional science signifies a greater open-mindedness. And it doesn't hurt to be a well-known scientist when you suggest such seemingly farfetched notions. For many people, the supposed authority of the authorities removes the element of crackpottedness from such schemes—which, of course, is a questionable proposition that implies that scientists cannot be crackpots. What science fiction writer, for instance, could get away with another scheme proposed by Dyson, the idea of colonizing comets? Dyson has said that trees could be grown on comets. Unlimited by gravity, they would stretch for hundreds of miles into space. And when men came to live on the comets, Dyson added, they would find themselves returning to the arboreal existence of their ancestors! While a serious science fiction writer would not publish such an idea, an association with reputable universities and other institutions now allows scientists to publish bad science fiction in the guise of science.

In any event, no signs of supercivilizations, or any other civilizations, have been detected so far.

The obvious question arises, if they are so smart, why haven't they contacted us?

Sagan has suggested that highly advanced civilizations may have discovered "hypothetical new laws of physics about which we can only dimly guess." The preferred SETI communications channels for Type II and Type III civilizations may lie within those still-unknown realms of physics.

Sagan has also wondered "if we are too early." Is it possible, he has asked, that intelligence arose in our solar system at a statistically unlikely, early moment? This question echoes the Canadian paleobiologist Dale Russell's suggestion that man might be the oldest intelligent creature in the cosmos. (See Chapter Eight, "Future Man.")

To Frank Drake, the reasons why extraterrestrial civilizations have not contacted us are clear. On the one hand, says Drake, it is true that the existence of Earth and other life-bearing planets probably can be established by many civilizations. After doing that, they may send space probes to the planets, but Drake figures that we might expect only one such spacecraft to arrive at Earth, at best, only every five thousand years or so. Says Drake: "They may have been here, or they may be on their way, but it is not surprising that there has been no such event in recorded history.

"If we detected in a nearby star system a planet much like the Earth," adds Drake, "we might send one spacecraft, but then no more unless it detected life at an interesting level. If other civilizations are like us, only a few spacecraft every few billion years might arrive at the Earth.

"The alternative is to send radio signals. There the answer to our lack of detection is simple: we have not listened nearly enough. In particular, we have not searched the radio spectrum significantly, and surely we have not searched enough of the sky. There could be easily detectable signals coming our way at all times, but we could have missed them. We simply have not tried hard enough."

Still another reason we may not have heard from anyone or haven't been visited is the Earth's out-of-the-way location, on the edge of our home galaxy's arm, far from its center.

Furthermore, societies in space may be so far advanced that they may have no more interest in communicating with us than we do with microbes.

The radio spectrum may be the wrong place to look for signals. Charles Townes, one of the inventors of the laser, has stressed, perhaps not surprisingly, the need to look in a variety of ways, including on laser frequencies. Says Townes: "The discovery may come in some very unexpected way through a completely different route."

Advanced civilizations, another scientist has speculated, may make up "a galactic Polynesian archipelago whose tenants are mostly concerned about their inner life and not anxious to communicate with beings on other planetary systems, and particularly not with us."

The most likely reason for the silence, though, could be a rarity, or even a complete absence, of high-technology civilizations. As we have seen throughout this book, evolution is much more likely to produce intelligent creatures that are unlike man than those that are very much like him. Trying to communicate with dolphins from the vicinity of another star obviously would produce no response.

If technical civilizations were abundant, certainly some of them would have investigated the Earth by now. We have seen how far and how fast the progress of science and technology can take us. In less than one hundred years, just one branch of science—physics—through its quest into the nature of matter has dramatically changed both the face of the Earth and our view of the universe. We have the SETI search, television, and transistor technology today because the question of what the atom consists of led to the discovery of the electron, only eighty years ago. The next question the physicists asked themselves, concerning what the nucleus of the atom is made of, led to the discovery of fission and fusion.

The continuing probing of both the subnuclear and of the cosmic realms is likely to lead to even more exciting discoveries. Huge atomic accelerators as well as new telescopes may yield new knowledge that will change our view of the universe as dramatically as Einstein's ideas did earlier in the century. "A small discovery in high-energy physics," says Robert R. Wilson, former director of the Fermi National Accelerator Laboratory, at Batavia, Illinois, "could change all the rules and produce technologies impossible to think of now." Many other scientists agree that the greatest discoveries about the universe, the nature of matter, of the human mind—and the interconnections among all three—are yet to come.

Are we to assume that advanced civilizations, if they exist, have not moved far beyond the still fairly primitive state of our science and technology?

Although such advanced science, as Arthur C. Clarke has noted, to us may be "indistinguishable from magic," there are limits, to be sure, to what another civilization could do. As far as we know, neither its signals nor its spaceships could travel faster than light—a strong argument against visitations by flying saucers and other types of UFOs.

We can't take seriously the UFO accounts as manifestations of extraterrestrial intelligences. For one thing, most of UFOs' reputed activities show more mischief than mind. For another thing, the idea that the Earth is being constantly visited by representatives of hundreds of civilizations—judging by the bewildering variety in the appearance of the alleged aliens—is hard to swallow. As Allan Hendry, chief investigator for the Center of UFO Studies, in Evanston, Illinois, concluded not long ago after studying a number of supposed abductions by aliens, the solution to such cases "is somehow intracranial rather than extraterrestrial." In other words, it's more likely to lie in the abductees' heads than among the stars. Hendry reached his conclusion after noticing a remarkable similarity in the accounts of abductions in the way the abductions supposedly occurred, and, at the same time, a great divergence in the details of individual abductions.

The behavior of the flying saucers and other UFOs, unchanged over the centuries, also hints at their natural origin—either in the minds of the observers or in the environment. Otherwise, if we are a dangerous culture, conspicuous, continual monitoring of the Earth that doesn't seem to lead to anything doesn't make sense. If we are not dangerous, proper contact would be in order. The kind of contact the flying-saucerians engage in, however, is irrational. They seem to seek out obscure people in obscure places and impart platitudes such as "We come to save you from yourselves." This doesn't sound like a manifestation of higher intelligence.

The UFO sightings can be explained as weather balloons, airplanes, satellites, the planet Venus, even reflection of sunlight from the wings of high-flying birds. Other UFOs undoubtedly can be explained by the new science of magnetohydrodynamics. It has shown, for instance, that the striking phenomenon of ball lightning, which appears to be responsible for a number of UFO reports, is in reality a strange ball of incandescent plasma. Unstable magnetic fields hold it together, making it perform erratic and unpredictable movements. It can appear and disappear suddenly. It can dart about, hover, or explode. It can perform many other maneuvers attributed to flying saucers. Appearing along the lines of electrical and magnetic fields that surround aircraft in flight, such blobs of plasma can execute the complex maneuvers that World War II pilots attributed to the mysterious luminescent "foo fighters" that sometimes accompanied their planes at close range. The blobs of plasma could "anticipate" the pilots' moves, dart off at strange angles, and show up on radar, to boot, to lend credibility to their appearance. On the ground, such concentrations of plasma can affect the electrical ignition systems of cars, to which they would be naturally attracted. High in the air, such blobs of plasma can assume the shapes of disks and similar saucerian forms.

It has occurred to some people that some of the UFOs may be holograms projected across interstellar distances. If so, they still look rather childish—as if some mischievous youngster on another planet was playing long-distance electronic games. We already have the capability to project holograms over short distances. This is accomplished with a laser shining through a specially prepared image of an object recorded on a high-resolution photographic plate. The laser shines through the plate, projecting the image of an object or a person as a ghostly three-dimensional copy literally hanging in the air. One such projection in a jewelry-store window on New York's Fifth Avenue of a disembodied woman's hand wearing a diamond ring and dangling a bracelet, was attacked, by an elderly woman passerby, with an umbrella; she termed the hologram "a work of the devil."

We can also discount some ufologists' suggestion that the UFOs are here because of the great leaps in our science and technology during and after World War II. Such a belief reflects an ignorance of both astronomy and technology. Our space-faring capabilities are so primitive that they would not have been noticed by outsiders. None of our spacecraft have yet left the solar system, and even after they do, it will take them hundreds of thousands of years to reach the nearest star.

The ufologists also sometimes suggest that the flashes of our atomic-bomb blasts have alerted the aliens. But scientists have calculated that even if the world's stockpile of nuclear weapons were transported to the far side of the moon and set off there, that flash would be seen by someone in space only if the observer was looking at the spot directly at that instant. Individual atom-bomb explosions are unlikely to be seen in interstellar space.

If, on the other hand, alien radio or other electromagnetic signals have reached the Earth in the past, there would have been no way for man to detect them. The stories of ancient astronauts do not sound plausible. The alleged markings of extraterrestrial visitations, such as the long white lines in the Nasca plains, of Peru, interpreted by adherents of visitations from the stars as "landing strips" for cosmic travelers, appear to be nothing more than religious or ritual drawings of birds, spiders, and other Earth-dwelling creatures. Carl Sagan has asked the reasonable question whether anyone could sensibly expect the extraterrestrials to prepare landing strips for B-29-type aircraft. No artifacts of visits from space have ever been found on Earth.

Could mysterious radio echoes be artifacts of that type? A series of such echoes had been received in Eindhoven, the Netherlands, and in Oslo during the 1920s and never deciphered. The echoes appeared to be bouncing off something in the solar system far beyond the moon. The radio astronomer Ronald Bracewell then suggested that a space probe from another star might have been circling the Earth, trying to make contact with us.

Enter Duncan Lunan, a young Scottish scientist. In the early 1970s, Lunan said he found that those signals could be interpreted as star maps, pointing out a particular star, Epsilon Boötis, and even supplying detailed information on the probe itself. Lunan and his associates claimed that they could show by plotting the echoes that the probe had transmitted a wealth of information on its planetary system. From these data, Lunan drew the conclusion that the senders of the probe had been desperately looking for a new planet as a home, because their main star had begun to turn into a fiery giant. Unfortunately, the alleged interstellar probe has not been heard from since the 1930s, when radio echoes were recorded once again.

While Lunan and his associates may have performed a valuable service in showing in great detail how subtle and indirect an interstellar contact may turn out to be, the idea that the Earth is being selected for special attention leaves many scientists unimpressed. They find the basic assumption, that we are so interesting as a race that other intelligent beings find it difficult to stay away from Earth, hard to believe. Scientists find the tendency of many people to accept the most complicated explanation for historical events and phenomena somewhat puzzling if not outright irrational. An example is the idea that astronauts from other civilizations landed on Earth in ancient times and instructed primitive men in how to build pyramids and in other arts and sciences. Similarly, those mysterious radio echoes were ascribed to a probe instead of to a much more likely natural source: plasma clouds, blown away from the sun, serving as signal reflectors.

"Some people have a need to fantasize," says Bob Edelson. "To imagine that someone is watching over us. The assumption that we are of irresistible interest to other civilizations smacks of arrogance. We may be of no more interest than the primitive tribes are to our anthropologists—and you don't see thousands of anthropologists descending on New Guinea."

Perhaps the signals from the stars are being transmitted in a manner not readily discernible to us.

An unusual type of interstellar message has been suggested by Japanese scientists. They have proposed that a virus known as Phix-174, which infects intestinal bacteria, might be itself such a message. The reason for this suggestion was the discovery that the sequence of the DNA bases in the virus seemed more artificial than natural. In this particular virus, it appeared that the DNA code could be read in three meaningful ways, an arrangement that seemed puzzling for a natural sequence. The investigators, however, were unable to construct a meaningful picture out of those DNA combinations.

Why would an advanced civilization resort to such a strange means of communication? The Japanese scientists argued that dispersion of viruses would be a more effective means than radio signalling since radio would work only if the receiving antennas happened to be pointed at the sender at the exactly correct time and tuned to the exact frequency. However, the Japanese scientists also admitted that there are obvious disadvantages to sending viruses. First, the particles would have to survive the long journey through interstellar space, where radiation could easily destroy them unless the viruses were somehow encapsulated. Second, the viruses would have to land on a biochemically suitable planet. Third, they would have to maintain a stable DNA sequence through the ages, somehow resisting mutations, which would alter the code, assuming there was one. A transposition of a single element in a binary message can render the whole message meaningless. All in all, the idea of viruses as messages seems rather farfetched. On the other hand, the assumption that the signal from the stars will come in as a mirror image of our type of radio transmission may be exceedingly naïve.

For our part, we have now moved beyond passive listening for signals. Messages have been beamed to the stars, and a number of spacecraft that will eventually leave the solar system and one day may fly by some stars, carry plaques and other paraphernalia for the edification of creatures of other stars.

Among the number of messages beamed to the stars is the one facing this page, created by Frank Drake and his associates and sent through the Arecibo radio telescope.

The Arecibo radio message is three minutes long, and it was beamed at a globular cluster of stars in the constellation Hercules, twenty-four thousand light-years away, on November 24, 1974. When the message reaches its destination, the beam will have spread to cover all the three hundred thousand stars in the Hercules cluster

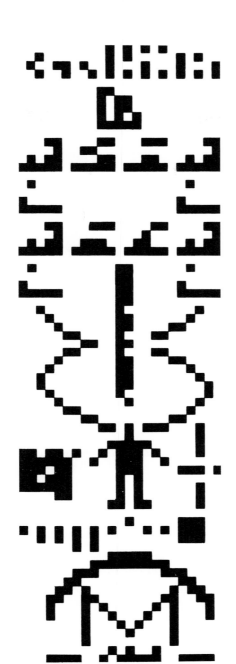

in the hope that some will be listening. But the response would not arrive on Earth until about A.D. 50,000.

The message itself consists of 1,679 binary pulses, which are represented by the zeros and ones of the computer code. An arrangement of the dots into seventy consecutive groups of twenty-three characters each would yield a visual image when the zeros are viewed as white space and the ones as black squares. What emerges on the top of the message is a listing of the numbers one to ten in binary form. This is the key to translation of the message.

In subsequent figures, the message displays the atomic numbers of hydrogen, carbon, nitrogen, oxygen, and phosphorus—the latter four being the basic constituents of life, and hydrogen being the most abundant element, which is also needed for life to arise and to continue. The message goes on to describe the building blocks of DNA and even shows the double-helix pattern of the heredity-determining DNA molecule.

Below the double helix, the message shows a rough outline of a human figure with a measure of its height and the size of the human population. Below the human figure is a representation of the nine planets of the solar system and of the sun, with the square representing the Earth indented toward the human figure's feet. Finally, the bottom of the message shows an outline of the Arecibo telescope beaming the message, and gives the telescope's size: 1,000 feet, or 330 meters.

As for spacecraft as messengers to the stars, NASA's Pioneers 10 and 11, launched in 1972 and 1973, are equipped with identical plaques designed to show when the spacecraft were launched, from where, and by the type of beings. The message was etched on a gold-anodized aluminum plate. It contained on its left side radiating lines representing the positions of fourteen pulsars arranged to indicate our sun as the home star of the civilization that had launched the craft. Two figures on the right-hand side of the plaque, of a man and a woman, symbolized the type of creature that had built the spacecraft. The man's right hand is raised in a gesture of good will. Across the bottom of the plaque are the planets. They range outward from the sun, with the spacecraft trajectory arching away from the Earth, passing Mars, and swinging by Jupiter.

Indicative of the vastness of the solar system, much less of the galaxy or the universe, those Pioneer spacecraft, whose primary mission to investigate Jupiter is now completed, are still inside the solar system and won't leave it until 1986. Traveling at 6 miles (9.5 kilometers) a second, the two craft would not reach the nearest stars for more than one hundred thousand years, even if they were heading in that direction, which they are not. Rocketry, therefore, does not appear especially promising for interstellar communication.

A more complicated and more detailed message is on its way to the stars aboard the spacecraft Voyager 1 and 2, launched in August and September of 1977. The two craft have now successfully photographed both Jupiter and Saturn, revealing the dramatic volcanic eruptions on the Jovian satellite Io as well as the presence of a nitrogen atmosphere on Saturn's moon Titan. Eventually, the two craft will leave the solar system and head toward the stars.

The messages placed aboard the Voyagers were put together by a committee of scientists and educators assembled by Carl Sagan. The final product of this work was a two-hour-long phonograph record called "Sounds of Earth." On it are inscribed greetings from Earth in various languages, samples of various types of music, and natural sounds of the surf, wind, and thunder, as well as the sounds of birds, whales, and other animals.

The phonograph records are enclosed in aluminum containers that include a porcelain cartridge, a diamond stylus, and playing instructions. Each recording begins with electronically encoded pictures, photographs of people in various countries, landscapes of Earth, various human activities, and views of the solar system. There is a message from then-President Jimmy Carter and one from UN Secretary General Kurt Waldheim, who expressed himself concisely:

> As the Secretary General of the United Nations, an organization of 147 member states who represent almost all of the human inhabitants of the planet Earth, I send greetings on behalf of the people of our planet. We step out of our solar system into the universe seeking only peace and friendship, to teach if we are called upon, to be taught if we are fortunate. We know full well that our planet and all its inhabitants are but a small part of the immense universe that surrounds us, and it is with humility and hope that we take this step.

Even if the record is not deciphered, the spacecraft itself, if retrieved by representatives of another civilization, would be a message from Earth, of course. Other imprints of man's presence are the spacecraft landed on Mars and Venus and those famous footprints on the moon, along with U.S. and Soviet spacecraft left there.

There is even a time capsule orbiting Earth. This is a satellite called Lageos, for Laser Geodynamic Satellite. Launched in May 1976, Lageos weighs 200 pounds (440 kilograms) and is a sphere made up of prism reflectors. A laser beam shined at the satellite from Earth makes Lageos reflect the light with exquisite precision, enabling scientists to study the tiny motions of continents as they continue to drift like huge rafts, as well as to measure the Earth's rotation and wobble about its axis. So stable is Lageos in its orbit that it should stay up there for eight million years.

If man's descendants or visitors from elsewhere ever retrieve Lageos, inside it they will find a small plaque that describes the satellite's mission and shows in three drawings the position of the continents when they all constituted the great world continent Pangaea (see illustration in Chapter Four, "The Turns and Twists of Evolution on Earth"), as well as their position now and where scientists expect them to be 10 million years from now.

All these attempts at contacting intelligent inhabitants of the cosmos may be misdirected, of course, since the life-forms elsewhere may be maintaining "radio silence" because they are not technological beings. There is nothing in our fossil record that would indicate an unbroken evolutionary progression from amoeba to man. What we are seeing instead is that intricate branching of the evolutionary tree, with a constant change in both direction and rate of evolution. "Man is the end of one ultimate twig," the noted American paleontologist George Gaylord Simpson has said. "The housefly, the dog flea are similarly the ends of theirs." In light of man's almost improbable appearance on Earth, the result of an unrepeatable sequence of events, man-like intelligent beings, may be so rare that contact with them may be all but impossible.

But discovery from afar, and eventual face-to-face encounters with life-forms such as the intelligent octopus, intelligent insect-like creatures, and brainy offspring of reptiles, as well as many similar beings, is not out of the question. Most of these creatures would probably have evolved vocal cords or some equivalent and developed some form of language so that we could communicate with them.

What might we learn from such creatures? Not the usual pap that radio astronomers speculate about. Nontechnological societies are not going to tell us how to cure cancer or to control thermonuclear arms. We may learn instead, to our sorrow and shame, what we should have learned from our own, not always humane history long before we started on the road to the stars: that the survival of an intelligent species requires not higher technology but greater humanity. To have that demonstrated to us by nonhuman creatures could be the ultimate insult to our vanity.

But there could also be a positive lesson in the discovery that man is alone in the universe, a unique, unrepeatable being. Such realization may also set in after a failure to locate any intelligent signals in space after, say, a one-hundred-year-long search. The knowledge that our kind of life is found only on one fragile planet could drive some sense into the heads of politicians as they tinker with their lethal arsenals. It may even keep their fingers off the buttons that would plunge the world into a nuclear war.

The big unanswered question is whether there is intelligent life on Earth, intelligent enough not to wipe the human race and most of the other life off the face of the Earth in the madness of a thermonuclear holocaust before discovering man's real worth and his place among the wonders of Darwin's universe.

INDEX

MP1M

QH 371 .B86 1981
Bylinsky, Gene.
Life in Darwin's universe